SPACE-TIME NUCLEAR REACTOR KINETICS

NUCLEAR SCIENCE AND TECHNOLOGY

A Series of Monographs and Textbooks

CONSULTING EDITOR

V. L. PARSEGIAN

School of Engineering
Rensselaer Polytechnic Institute
Troy, New York

---

1. John F. Flagg (Ed.)
   CHEMICAL PROCESSING OF REACTOR FUELS, 1961
2. M. L. Yeater (Ed.)
   NEUTRON PHYSICS, 1962
3. Melville Clark, Jr., and Kent F. Hansen
   NUMERICAL METHODS OF REACTOR ANALYSIS, 1964
4. James W. Haffner
   RADIATION AND SHIELDING IN SPACE, 1967
5. Weston M. Stacey, Jr.
   SPACE-TIME NUCLEAR REACTOR KINETICS, 1969

Space-Time Nuclear Reactor Kinetics

WESTON M. STACEY, JR.

REACTOR PHYSICS DIVISION
ARGONNE NATIONAL LABORATORY
ARGONNE, ILLINOIS

1969

ACADEMIC PRESS New York and London

ACADEMIC PRESS, INC.
111 Fifth Avenue, New York, New York 10003

United Kingdom Edition published by
ACADEMIC PRESS, INC. (LONDON) LTD.
Berkeley Square House, London W1X 6BA

LIBRARY OF CONGRESS CATALOG CARD NUMBER: 70-84158

PRINTED IN THE UNITED STATES OF AMERICA

CONTENTS

CHAPTER 3. Variational Synthesis Methods

CHAPTER 4. Stochastic Kinetics

CHAPTER 5. Xenon Spatial Oscillations

CHAPTER 6. Stability

CHAPTER 7. Control

SPACE-TIME
NUCLEAR REACTOR KINETICS

INTRODUCTION

Until quite recently, nuclear reactor kinetic theory was based almost entirely upon the "point kinetics" model, wherein the neutron flux distribution is assumed to be invariant in space with a time varying amplitude. This theory served quite well in the days of small, tightly coupled reactors in which the spatial neutron distribution was relatively insensitive to local changes in material properties.

However, the point kinetics model, in its unadorned state, has been found drastically inappropriate for transient analyses in the larger reactors being considered for economically competitive power production. In addition to errors in the prediction of the total power arising from the use of incorrect spatial neutron flux distributions in obtaining average reactor properties, the point kinetics model is intrinsically unable to describe purely spatial neutron flux tilting such as may be induced by an oscillating xenon spatial distribution. The need for more sophisticated methods, together with a dramatic increase in digital computing capability which renders the application of more elaborate methods feasible, has given impetus to the development of a new reactor kinetic theory which accounts explicitly for the time varying spatial neutron flux distribution.

This book is devoted to recently developed theory in those areas which collectively constitute the subject of space-time nuclear reactor kinetics. Related topics that are adequately treated elsewhere, such as the treatment of feedback mechanisms and safety considerations in different reactor types,[1] methods for solving the point kinetics equations[2,3] and experimental methods and the source of basic data[4] are omitted. Emphasis is upon the basic theory and computational methods.

The first three chapters are concerned with computational methods for solving the spatially dependent neutron kinetics equations. A mathematical development of finite-difference, modal expansion (synthesis), and nodal approximations is the substance of Chapter 1. The point kinetics approximation is developed within the same framework, and extensions to incorporate spatial flux changes, such as the adiabatic and quasistatic approximations, are presented. The physical and mathematical properties of the various approximations are discussed.

Explicit and implicit numerical integration algorithms for the approximations of Chapter 1 are compared in Chapter 2. Properties of the various algorithms, such as stability, truncation error, and computational effort, are discussed. The discussion is limited to those methods that have been investigated for the spatially dependent neutron kinetics approximations (forward and backward differences, θ-difference, time-integrated, perturbation expansion, GAKIN, and ADI).

Variational techniques, and their application in the development of space-time flux synthesis approximations, are reviewed in Chapter 3. A general variational formulation of the neutron kinetics problem is presented and used in the development of multichannel flux synthesis equations.

The next two chapters treat physical phenomena that are not adequately covered in other books. For this reason, these chapters have more of a physical flavor than the previous chapters, although computational methods are still stressed.

The basic neutronic processes that occur in a nuclear reactor are stochastic in nature. Kinetic theories for these processes are described in Chapter 4. The deterministic theory upon which the other chapters (and the vast majority of nuclear reactor physics) are based is shown to describe the mean value of the stochastic distribution function. Certain general properties of the stochastic distribution are inferred from numerical results. In addition to its basic theoretical value, the stochastic theory is useful in the analysis of power fluctuations in operating reactors and in the analysis of weak-source startups.

Xenon spatial oscillations are discussed in Chapter 5. The necessity to control or suppress these oscillations in the large thermal power reactors presently being designed has led to considerable activity in this area recently. Linear stability criteria used in predicting the inception of such oscillations are developed and compared with experiments and numerical simulations. An attempt was made to select from the sizeable literature on the subject those linear stability analyses that could be implemented readily for the spatially detailed cores presently being designed. Effects of nonlinearities and control rod motion on xenon oscillation analyses is illustrated by numerical simulation. Control of xenon spatial oscillations is discussed, and the application of dynamic programming to this problem is described.

Chapters 6 and 7 treat the stability and control, respectively, of spatially dependent reactor models. These topics are not so well developed at this time as is the subject matter of the first five chapters. Consequently, the intention of these last two chapters is to present what seems to be the appropriate basic theory, in hope that it will stimulate the necessary extensions and applications.

Space-time nuclear reactor kinetics is today an area of active and rapid development. Although the methods discussed in this book may well be supplanted by more ingenious techniques in the future, a large number of people are currently involved in trying to understand, implement, and extend the present theory and methods. This interest is motivated not only by the intellectual challenge involved, but primarily by the very practical requirements for predicting the performance of proposed reactor designs. Thus, it is hoped, by reviewing in some detail the current status of space-time nuclear reactor kinetics, that this book will serve a useful purpose in bridging the gap between the older texts on reactor kinetics and the new work appearing in the current technical literature.

REFERENCES

1. T. J. Thompson and J. G. Beckerly (eds.), *The Technology of Nuclear Reactor Safety*, Vol. I. M.I.T. Press, Cambridge, Massachusetts, 1964.
2. M. Ash, *Nuclear Reactor Kinetics*. McGraw-Hill, New York, 1965.
3. L. Shotkin, "Mathematical Methods in Reactor Dynamics." Academic Press, New York (in preparation).
4. G. R. Keepin, *Physics of Nuclear Kinetics*. Addison-Wesley, Reading, Massachusetts, 1965.

Chapter 1

SPATIAL APPROXIMATIONS

1.1 The Time-Dependent Group-Diffusion Equations

Although a rigorous description of nuclear reactor kinetics would invoke energy dependent neutron transport theory, or some high order approximation thereto, it is unlikely that problems in reactor kinetics will, in the forseeable future, receive a more rigorous treatment than that represented by the group-diffusion approximation which is widely used in reactor statics. The time-dependent group-diffusion equations describe the average reaction rate over an interval of energy referred to as a group in terms of neutron-diffusion theory, and have the generic form

$$\begin{aligned}
&\nabla \cdot D^g(r, t)\, \nabla\phi^g(r, t) - (\Sigma_a{}^g(r, t) + \Sigma_s{}^g(r, t))\phi^g(r, t) \\
&\quad + \sum_{g' \neq g}^{G} \Sigma_s^{g'/g}(r, t)\phi^{g'}(r, t) + (1 - \beta)\chi_P{}^g \sum_{g'=1}^{G} \nu^{g'}\Sigma_f{}^{g'}(r, t)\phi^{g'}(r, t) \\
&\quad + \sum_{m=1}^{M} \lambda_m \chi_m{}^g C_m(r, t) + Q^g(r, t) = (1/v^g)\dot{\phi}^g(r, t), \qquad (1.1) \\
&\qquad\qquad\qquad\qquad g = 1, \ldots, G.
\end{aligned}$$

The quantity D^g is the diffusion coefficient in group g, while $\Sigma_a{}^g$, $\Sigma_s{}^g$, and $\Sigma_f{}^g$ represent the macroscopic absorption, scattering removal, and fission cross sections, respectively, in group g. The probability per unit time per unit neutron flux that a neutron in group g' is scattered into group g is denoted by $\Sigma_s^{g'/g}$. The quantity $\nu^{g'}$ is the average number of neutrons produced in a fission induced by a neutron in group g', and v^g is the average speed of neutrons in group g. The neutron flux in group g is denoted by ϕ^g, and Q^g is the external source of neutrons in group g. The quantity C_m represents the concentration of delayed neutron precursors of type m, with decay constant λ_m and delay fraction $\beta_m (\beta \equiv \Sigma_{m=1}^{M} \beta_m)$. The fraction of the neutrons produced directly by fission and by precursor decay that have energy within group g are denoted $\chi_P{}^g$ and $\chi_m{}^g$, respectively.

The precursors satisfy a simple balance equation

$$\beta_m \sum_{g=1}^{G} \nu^g \Sigma_f{}^g(r, t)\phi^g(r, t) - \lambda_m C_m(r, t) = \dot{C}_m(r, t), \qquad (1.2)$$
$$m = 1, \ldots, M.$$

Associated with Equations (1.1) and (1.2) are boundary conditions of the general form

$$\phi^g(R, t) + d\,\nabla\phi^g(R, t) = 0,$$

where R denotes the external boundary of the reactor and d is an extrapolation distance. At internal interfaces, flux and current continuity conditions must be satisfied; i.e.,

$$\phi^g(r_+, t) = \phi^g(r_-, t),$$
$$D^g(r_+, t)\,\nabla\phi^g(r_+, t) = D^g(r_-, t)\,\nabla\phi^g(r_-, t).$$

In most problems, initial conditions of the form

$$\phi^g(r, 0) = g^g(r),$$
$$C_m(r, 0) = h_m(r),$$

will be specified.

Discussions and derivations of the multigroup diffusion equations may be found in the books by Weinberg and Wigner[1] and Davison,[2] and definitions of the group constants are given, among other places, in Section 7.1.1 of Reference 3. A brief derivation is given in the Appendix for convenience. Data and methods for determining group constants are also given in Reference 3, although more recent data are incorporated in the Evaluated Nuclear Data Files (ENDF). Habetler and Martino[4] present the mathematical theory of the multigroup diffusion model.

Total delayed neutron yields for the principle fissile and fertile nuclei are given in Table 1.1, and precursor yields and decay constants for ^{235}U are given in Table 1.2. Additional delayed neutron data, including photoneutron data, are given by Keepin.[5]

TABLE 1.1

ABSOLUTE TOTAL DELAYED NEUTRON YIELDS

Nuclide	Delayed neutrons/fission ($\beta\nu$)	
	Fast fission	Thermal fission
^{239}Pu	0.0063	0.0061
^{233}U	0.0070	0.0066
^{240}Pu	0.0088	—
^{241}Pu	—	0.0154
^{235}U	0.0165	0.0158
^{238}U	0.0412	—
^{232}Th	0.0496	—

TABLE 1.2

DELAYED NEUTRON PARAMETERS FOR THE THERMAL FISSION OF URANIUM-235

Precursor type, m	Decay constant λ (sec^{-1})	Relative yield, β_m/β	Absolute yield, $\beta\nu$
1	0.0124	0.033	0.052
2	0.0305	0.219	0.346
3	0.111	0.196	0.310
4	0.301	0.395	0.624
5	1.14	0.115	0.182
6	3.01	0.042	0.066

Equations (1.1) and (1.2) are coupled partial and ordinary, respectively, differential equations, that can be solved analytically only for the simplest cases. This chapter, and the next two, are devoted to approximations that reduce these equations to a form amenable to computation. In this chapter, approximations that reduce these equations to coupled ordinary differential equations in the time variable are considered. Chapter 2 is concerned with the solution of these coupled ordinary differential equations by numerical integration methods. In Chapter 3, approximations for the spatial dependence and numerical integration methods for the time dependence are derived simultaneously from a variational principle.

It is convenient to write the G Equations (1.1) as a matrix equation

$$(\nabla \cdot \mathbf{D} \nabla - \mathbf{R}_a - \mathbf{R}_s + \mathbf{S} + (1-\beta)\chi_P \mathbf{F}^T)\phi + \sum_{m=1}^{M} \lambda_m \chi_m C_m + \mathbf{Q} = \mathbf{V}^{-1}\dot{\phi}, \tag{1.3}$$

where $\boldsymbol{\phi}$ is a $G \times 1$ column vector of group fluxes, χ_P and χ_m are $G \times 1$ column vectors of prompt fission and precursor decay neutron spectra, and $\mathbf{F}^T$ is a $1 \times G$ row vector with elements $\nu^g \Sigma_f{}^g$. $\mathbf{Q}$ is a $G \times 1$ column vector of group sources. The $G \times G$ matrices $\mathbf{D}$, $\mathbf{R}_a$, $\mathbf{R}_s$, and $\mathbf{V}^{-1}$ are diagonal with elements D^g, $\Sigma_a{}^g$, $\Sigma_s{}^g$, and $1/v^g$, respectively, while $\mathbf{S}$ is a $G \times G$ matrix with elements $\Sigma_s^{g/g'}$.

Equations (1.2) can also be written in matrix notation

$$\beta_m \mathbf{F}^T \phi - \lambda_m C_m = \dot{C}_m, \qquad m = 1, \ldots, M. \tag{1.4}$$

1.2 The Spatial Finite-Difference Approximation

The spatial finite-difference approximation has been employed widely to obtain solutions to the static group-diffusion equations, and has been applied successfully to the time-dependent group diffusion equations in a limited number of cases.

To obtain the finite-difference approximation, the reactor model is partitioned into a finite number of elemental regions, with each region enclosing a "mesh point." Equations (1.1) and (1.2) are then integrated over each elemental region, with some numerical approximations, to obtain a coupled set of ordinary time-dependent differential equations in the unknown group fluxes and precursor densities at the mesh points. Thus, the requirement that Equations (1.3) and (1.4) be satisfied throughout the spatial domain is replaced by the requirement that these equations be satisfied in an integral sense over each of the elemental regions.

One technique for obtaining finite-difference equations is known as the box integration method,† in which the elemental region associated with each mesh point is a box whose surfaces bisect the distance between the mesh point in question and its nearest neighbors, and in which the nuclear properties change only at mesh points. With reference to the inset figure, one-dimensional finite-difference equations are derived by integrating Equations (1.1) from $X_i - \Delta_i/2$ to $X_i + \Delta_{i+1}/2$, and making some numerical approximations.

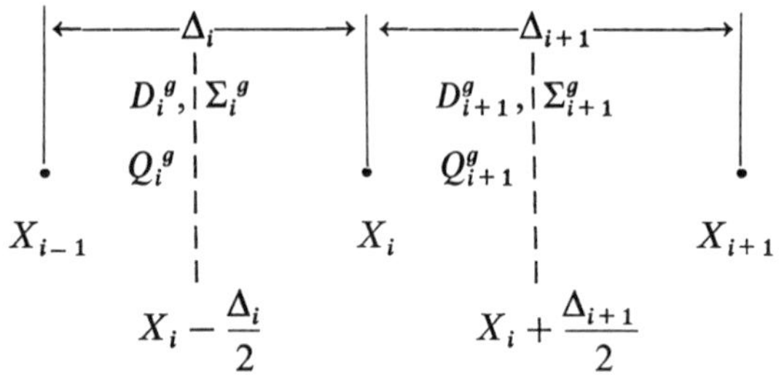

The diffusion term is of the form

$$\int_{X_i-(\Delta_i/2)}^{X_i+(\Delta_{i+1}/2)} dX\, \nabla \cdot D^g(X)\, \nabla\phi^g(X)$$

$$= D_i^g \int_{X_i-(\Delta_i/2)}^{X_i} dX\, \nabla^2\phi^g(X) + D_{i+1}^g \int_{X_i}^{X_i+(\Delta_{i+1}/2)} dX\, \nabla^2\phi^g(X)$$

$$= D_i^g\, \nabla\phi^g \Big|_{X_i-(\Delta_i/2)}^{X_i} + D_{i+1}^g\, \nabla\phi^g \Big|_{X_i}^{X_i+(\Delta_{i+1}/2)}$$

$$\cong D_i^g \frac{(\phi_{i-1} - \phi_i)}{\Delta_i} + D_{i+1}^g \frac{(\phi_{i+1} - \phi_i)}{\Delta_{i+1}},$$

where the continuity of current at X_i has been employed to cancel derivatives evaluated at X_i, and central difference approximations have been made for derivatives evaluated at $X_i - \Delta_i/2$ and $X_i + \Delta_{i+1}/2$. The non-diffusion terms are of the form

$$\int_{X_i-(\Delta_i/2)}^{X_i+(\Delta_{i+1}/2)} dX\, \Sigma^g(X)\phi^g(X) \cong \Sigma_i^g\, \phi_i^g\, \frac{\Delta_i}{2} + \Sigma_{i+1}^g\, \phi_i^g\, \frac{\Delta_{i+1}}{2}.$$

ϕ_i^g is the value of the flux at mesh point i.

† Development of finite-difference equations from the box integration method is discussed in detail by Wachpress,[6] pp. 70–74.

Finite-difference equations may be derived by the box integration method for a variety of two- and three-dimensional geometries. For N mesh points, the $N \times N$ matrix that represents $-\nabla \cdot D^g \nabla$, $\mathscr{D}^g$, is real, symmetric, and diagonal dominant, with nonpositive off-diagonal elements and positive diagonal elements. The $N \times N$ matrices that represent $\Sigma_a{}^g$, $\Sigma_s{}^g$, $\Sigma_s^{g'/g}$, $\nu^g\Sigma_f{}^g$, and v^g are real, diagonal matrices $\mathscr{R}_a{}^g$, $\mathscr{R}_s{}^g$, $\mathscr{S}^{g'/g}$, $\mathscr{F}^g$, and $\mathscr{V}^g$, respectively. The group fluxes and precursors are represented by $N \times 1$ column vectors $\mathbf{\Psi}^g$ and $\mathbf{d}_m$, respectively, and the group source is represented by the $N \times 1$ column vector $\hat{\mathbf{Q}}^g$, whose elements are

$$\hat{Q}_i^g = Q_i \frac{\Delta_i}{2} + Q_{i+1}^g \frac{\Delta_{i+1}}{2}.$$

The finite-difference approximation to the kinetic group-diffusion equations may be written as supermatrix equations,

$$\mathscr{V}[-(\mathscr{D} + \mathscr{R}_a) - \mathscr{R}_s + \mathscr{S} + (1 - \beta)\chi_P'\mathscr{F}^T]\mathbf{\Psi} + \sum_{m=1}^{M} \lambda_m \mathscr{V} \chi_m' \mathbf{d}_m + \mathscr{V}\hat{\mathbf{Q}} = \dot{\mathbf{\Psi}}, \tag{1.5}$$

$$\beta_m \mathscr{F}^T \mathbf{\Psi} - \lambda_m \mathbf{d}_m = \dot{\mathbf{d}}_m, \quad m = 1, \ldots, M, \tag{1.6}$$

where the supermatrices are defined

$$\mathscr{D} + \mathscr{R}_a \equiv \begin{bmatrix} (\mathscr{D}^1 + \mathscr{R}_a{}^1) & & 0 \\ & \ddots & \\ 0 & & (\mathscr{D}^G + \mathscr{R}_a{}^G) \end{bmatrix}$$

$$\mathscr{R}_s - \mathscr{S} \equiv \begin{bmatrix} \mathscr{R}_s{}^1 & -\mathscr{S}^{2/1} & \cdots & -\mathscr{S}^{G/1} \\ \vdots & \ddots & & \vdots \\ -\mathscr{S}^{1/G} & \cdots & & \mathscr{R}_s{}^G \end{bmatrix}$$

$$\mathscr{V} \equiv \begin{bmatrix} \mathscr{V}^1 & & 0 \\ & \ddots & \\ 0 & & \mathscr{V}^G \end{bmatrix},$$

$$\mathcal{F}^T \equiv \begin{bmatrix} \mathcal{F}^1 & \dots & \mathcal{F}^G \end{bmatrix},$$

$$\chi_P' \equiv \begin{bmatrix} \chi_P{}^1\mathbf{I} \\ \vdots \\ \chi_P{}^G\mathbf{I} \end{bmatrix}, \quad \chi_m' \equiv \begin{bmatrix} \chi_m{}^1\mathbf{I} \\ \vdots \\ \chi_m{}^G\mathbf{I} \end{bmatrix},$$

$$\boldsymbol{\Psi} \equiv \begin{bmatrix} \boldsymbol{\Psi}^1 \\ \vdots \\ \boldsymbol{\Psi}^G \end{bmatrix},$$

where $\mathbf{I}$ is an $N \times N$ identity matrix.

Equations (1.5) and (1.6) may be written compactly as

$$\dot{\boldsymbol{\theta}} = \mathbf{K}\boldsymbol{\theta} + \mathbf{P}, \tag{1.7}$$

where

$$\boldsymbol{\theta} \equiv \begin{bmatrix} \boldsymbol{\Psi} \\ \mathbf{d}_1 \\ \vdots \\ \mathbf{d}_M \end{bmatrix}, \quad \mathbf{P} \equiv \begin{bmatrix} \mathscr{V}\hat{\mathbf{Q}} \\ \mathbf{0} \\ \vdots \\ \mathbf{0} \end{bmatrix}, \tag{1.8}$$

$$\mathbf{K} \equiv \begin{bmatrix} \mathscr{V}\{(1-\beta)\chi_P'\mathcal{F}^T - \mathcal{L}\} & \lambda_1\mathscr{V}\chi_1' \cdots \lambda_M\mathscr{V}\chi_M' \\ \beta_1\mathcal{F}^T & -\lambda_1\mathbf{I} \quad \cdots \quad 0 \\ \vdots & \vdots \quad \ddots \quad \vdots \\ \beta_M\mathcal{F}^T & 0 \quad \cdots \quad -\lambda_M\mathbf{I} \end{bmatrix}, \tag{1.9}$$

and

$$\mathcal{L} \equiv \mathcal{D} + \mathcal{R}_a + \mathcal{R}_s - \mathcal{S}. \tag{1.10}$$

Because the total scattering out of group g must equal the sum of the scattering from group g to all other groups,

$$\mathcal{R}_s{}^g = \sum_{g' \neq g}^{G} \mathcal{S}^{g/g'}. \tag{1.11}$$

Thus, the matrix $\mathcal{L}$ of Equation (1.10) is essentially nonpositive with positive diagonal elements and column diagonal dominance.

The properties of the matrix **K** determine the nature of the solution of Equation (1.7) and influence the efficacy of numerical integration methods for the time dependence that are necessary to obtain a solution (see Chapter 2). From the foregoing discussion, it follows that **K** is essentially nonnegative and real, and thus has a nonnegative eigenvector.[7]

If each of the N elemental regions contains a fissionable species, and if neutrons can be introduced into each energy group by fission, precursor decay, or scattering, then **K** is an irreducible matrix. A corollary to the theorem of Frobenius (see Chapter 2 of Wachpress[6] and Section 8.2 of Varga[8]) states that a real, essentially nonnegative, irreducible matrix has a unique nonnegative eigenvector, which in fact is positive, with a real, simple eigenvalue ω_1 that is greater than the real part of any other eigenvalue of the matrix.† Moreover, the eigenvalue ω_1 increases when any element of the matrix increases. It follows that the transpose matrix, $\mathbf{K}^T$, is real, essentially nonnegative, and irreducible, and that $\mathbf{K}^T$ and **K** have the same eigenvalue spectrum.

Then, if $\boldsymbol{\theta}(0)$ is a nonnegative column vector, and if the elements of **K** are time independent, the source-free case of Equation (1.7) has a solution‡

$$\boldsymbol{\theta}(t) = \frac{\langle \boldsymbol{\Theta}^* \mid \boldsymbol{\theta}(0) \rangle}{\langle \boldsymbol{\Theta}^* \mid \boldsymbol{\Theta} \rangle}[\exp(\omega_1 t)]\,\boldsymbol{\Theta} + O[\exp(\omega_2 t)] \tag{1.12}$$

where $\boldsymbol{\Theta}$ and $\boldsymbol{\Theta}^*$ are the unique positive eigenvectors of **K** and $\mathbf{K}^T$, respectively, and

$$\mathrm{Re}(\omega_2) < \omega_1 . \tag{1.13}$$

The solution to Equation (1.7) with a nonzero source has the same time dependence.

Numerical studies[10, 11] of Equation (1.12) indicate that the eigenvalues of **K** occur in clusters of $M + G$, with all eigenvectors of a cluster having similar spatial shapes. The assumption that all eigenvectors in a cluster have identical spatial shapes leads to some useful orthogonality properties.[10] The M algebraically largest eigenvalues in a cluster have magnitudes that are of the order of the precursor decay constants; the G remaining

† The same type of positivity theorem can be proven under less restrictive, but more mathematical conditions.[9]

‡ The Dirac notation $\langle \mid \rangle$ is used to indicate a summation over discrete independent variables or an integral over continuous independent variables, as appropriate.

eigenvalues have larger magnitudes. For a critical reactor, algebraically the largest eigenvalue is zero.

When $G = 1$ (i.e., one-group theory), Equation (1.6) can be premultiplied by $\mathscr{V}$, and the source-free version of Equations (1.5) and (1.6) can be written,

$$\dot{\theta}' = \mathbf{K}'\theta', \tag{1.14}$$

where

$$\theta' \equiv \begin{bmatrix} \Psi \\ \xi_1 \\ \vdots \\ \xi_M \end{bmatrix}, \tag{1.15}$$

$$\mathbf{K}' \equiv \begin{bmatrix} \Lambda^{-1}(\mathbf{R}-\mathbf{B}) & \lambda_1\mathbf{I} \cdots \lambda_M\mathbf{I} \\ \Lambda^{-1}\mathbf{B}_1 & -\lambda_1\mathbf{I} \cdots \quad 0 \\ \vdots & \ddots \quad \vdots \\ \Lambda^{-1} \quad {}_M & 0 \quad \cdots -\lambda_M\mathbf{I} \end{bmatrix} \tag{1.16}$$

and

$$\begin{aligned} \mathbf{R} &\equiv \mathscr{F} - (\mathscr{D} + \mathscr{R}_a), \\ \mathbf{B}_m &= \beta_m \mathscr{F}, \\ \mathbf{B} &\equiv \sum_{m=1}^{M} \mathbf{B}_m, \\ \xi_m &\equiv \mathscr{V}\mathbf{d}_m, \\ \Lambda^{-1} &\equiv \mathscr{V}. \end{aligned} \tag{1.17}$$

The matrices Λ, $\mathbf{B}_m$, and $\mathbf{B}$ are positive definite, and the matrix $\mathbf{R}$ is symmetric. Thus, $\mathbf{K}'$ is a "kinetics matrix," and can be shown to have the following properties[12]:

(a) The values $\omega = -\lambda_1, \ldots, -\lambda_M$ are excluded from the spectrum of $\mathbf{K}'$.

(b) The eigenvalues of $\mathbf{K}'$ satisfy the secular equation,

$$\det\left(\omega\Lambda + \sum_{m=1}^{M} \frac{\beta_m\omega}{\lambda_m + \omega}\mathbf{B}_m - \mathbf{R}\right) = 0. \tag{1.18}$$

(c) The eigenvalues of $\mathbf{K}'$ are real, and $\mathbf{K}'$ has a complete set of eigenvectors.

(d) The solution of Equation (1.14), for the elements of $\mathbf{K}'$ time independent, is a sum of $N(M+1)$ real exponentials.

(e) Zero is a k-fold eigenvalue of $\mathbf{K}'$ if, and only if, it is a k-fold eigenvalue of $\mathbf{R}$.

(f) The intervals $(-\infty, -\lambda_1)$, $(-\lambda_1, -\lambda_2), \ldots, (-\lambda_M, +\infty)$, where $\lambda_1 > \lambda_2 > \ldots > \lambda_M$, each contain exactly N eigenvalues of $\mathbf{K}'$.

(g) If $\mathbf{R}$ has p positive and q negative eigenvalues, then $\mathbf{K}'$ has p positive and q negative eigenvalues in the interval $(-\lambda_M, +\infty)$.

(h) All eigenvalues of $\mathbf{K}'$ in the interval $(-\lambda_M, +\infty)$ are in the subinterval $(-\lambda_M, 0)$ when $k_{\text{eff}} \leqq 1$, and are in the subinterval $(0, +\infty)$ when $k_{\text{eff}} \geqq 1$.

1.3 Modal Expansion Approximations

Modal expansion techniques may be characterized as those procedures by which the number of independent variables in a problem is reduced by means of an expansion in known functions of one or more of the independent variables. The general theory of this technique has been discussed in some detail.[13–15] In this section, two types of modal expansion methods that have been employed in reactor kinetics will be described and different types of expansion functions will be discussed.

The "time-synthesis" approximation is derived by expanding the neutron flux in known functions $\varphi_n{}^g(x, y, z)$, with unknown expansion coefficients $a_n{}^g(t)$:

$$\begin{bmatrix} \phi^1(x, y, z, t) \\ \vdots \\ \phi^G(x, y, z, t) \end{bmatrix} \cong \sum_{n=1}^{N} \begin{bmatrix} \varphi_n{}^1(x, y, z) & 0 \cdots 0 \\ \vdots & \ddots \quad \vdots \\ 0 \cdots & \varphi_n{}^G(x, y, z) \end{bmatrix} \begin{bmatrix} a_n{}^1(t) \\ \vdots \\ a_n{}^G(t) \end{bmatrix}, \tag{1.19}$$

or

$$\boldsymbol{\phi}(x, y, z, t) \cong \sum_{n=1}^{N} \boldsymbol{\varphi}_n(x, y, z)\mathbf{a}_n(t).$$

The approximate equality is used because it is not, in general, possible to satisfy Equations (1.3) and (1.4) exactly with a solution of the form of relation (1.19).

The expansion indicated by relation (1.19) is substituted into Equations (1.3) and (1.4), the former is premultiplied by $\mathbf{W}_l$ and the latter is premultiplied by $\mathbf{W}_l\chi_m$, then both are integrated over the volume of the reactor. The matrix $\mathbf{W}_l$

$$\mathbf{W}_l(x, y, z) = \begin{bmatrix} W_l^1(x, y, z) & 0 & \cdots & 0 \\ 0 & \ddots & & \vdots \\ \vdots & & \ddots & \vdots \\ 0 & \cdots & & W_l^G(x, y, z) \end{bmatrix}, \tag{1.20}$$

consists of group weighting functions. This procedure is repeated for N different weighting functions, resulting in the set of equations

$$\sum_{n=1}^{N} [-\mathbf{A}_{ln}(t)\mathbf{a}_n(t) + (1-\beta)\mathbf{F}_{ln}^T(t)\mathbf{a}_n(t)] + \sum_{m=1}^{M} \lambda_m \mathbf{C}_{m,l}(t) + \mathbf{Q}_l(t) = \sum_{n=1}^{N} \tau_{ln}\dot{\mathbf{a}}_n(t), \tag{1.21}$$

$$l = 1, \ldots, N,$$

$$\beta_m \sum_{n=1}^{N} \mathbf{F}_{m,ln}^T(t)\mathbf{a}_n(t) - \lambda_m \mathbf{C}_{m,l}(t) = \dot{\mathbf{C}}_{m,l}(t), \tag{1.22}$$

$$m = 1, \ldots, M; \quad l = 1, \ldots, N,$$

where

$$\mathbf{A}_{ln}(t) \equiv \int dx\, dy\, dz\, \mathbf{W}_l[\nabla \cdot \mathbf{D}\,\nabla - \mathbf{R}_a - \mathbf{R}_s + \mathbf{S}]\boldsymbol{\varphi}_n, \tag{1.23}$$

$$\mathbf{F}_{ln}^T(t) \equiv \int dx\, dy\, dz\, \mathbf{W}_l \chi_P \mathbf{F}^T \boldsymbol{\varphi}_n, \tag{1.24}$$

$$\mathbf{F}_{m,ln}^T(t) \equiv \int dx\, dy\, dz\, \mathbf{W}_l \chi_m \mathbf{F}^T \boldsymbol{\varphi}_n, \tag{1.25}$$

$$\boldsymbol{\tau}_{ln} \equiv \int dx\, dy\, dz\, \mathbf{W}_l \mathbf{V}^{-1} \boldsymbol{\varphi}_n, \tag{1.26}$$

$$\mathbf{C}_{m,l}(t) \equiv \int dx\, dy\, dz\, \mathbf{W}_l \chi_m C_m, \tag{1.27}$$

$$\mathbf{Q}_l(t) \equiv \int dx\, dy\, dz\, \mathbf{W}_l \mathbf{Q}. \tag{1.28}$$

Equations (1.21) and (1.22) are coupled ordinary differential equations, which are solved for the $\mathbf{a}_n$ and $\mathbf{C}_{m,l}$. Then the neutron flux is constructed from relation (1.19). The physical significance of Equations (1.21) and (1.22) follows from the mathematical manipulations involved in their derivation. Equations (1.21) and (1.22) are requirements that a solution of the form of relation (1.19) satisfies Equations (1.3) and (1.4) in a weighted integral sense for N different weighting functions, the integral being taken over the entire spatial domain. Thus, the requirement that the neutron and precursor balance equations be satisfied at all spatial locations has been relaxed to require that these balance equations be satisfied in a weighted integral sense over the spatial domain for N different weighting functions.

The other general type of modal expansion method to be discussed in this section is "space-time" synthesis. The equations are derived by expanding the neutron flux in known functions $\varphi_n{}^g(x, y)$ of two spatial dimensions with unknown combining coefficients $a_n{}^g(z, t)$ which depend on the remaining spatial dimension and time. The same "substitute, weight, and integrate" procedure just described is employed to obtain the space-time synthesis equations, except that the weighting functions and the integration involve only the x and y coordinates:

$$\sum_{n=1}^{N}\left[\frac{\partial}{\partial z}\mathbf{D}_{ln}(z,t)\frac{\partial}{\partial z}\mathbf{a}_n(z,t)-\mathbf{A}_{ln}(z,t)\mathbf{a}_n(z,t)+(1-\beta)\mathbf{F}_{ln}^T(z,t)\mathbf{a}_n(z,t)\right]$$

$$+\sum_{m=1}^{M}\lambda_m\mathbf{C}_{m,l}(z,t)+\mathbf{Q}_l(z,t)=\sum_{n=1}^{N}\boldsymbol{\tau}_{ln}(z)\dot{\mathbf{a}}_n(z,t), \tag{1.29}$$

$$l=1,\ldots,N,$$

$$\beta_m\sum_{n=1}^{N}\mathbf{F}_{m,ln}^T(z,t)\mathbf{a}_{ln}(z,t)-\lambda_m\mathbf{C}_{m,l}(z,t)=\dot{\mathbf{C}}_{m,l}(z,t), \tag{1.30}$$

$$l=1,\ldots,N;\quad m=1,\ldots,M,$$

where

$$\mathbf{D}_{ln}(z,t)\equiv\int dx\,dy\,\mathbf{W}_l\mathbf{D}\boldsymbol{\varphi}_n, \tag{1.31}$$

$$\mathbf{A}_{ln}(z,t)\equiv\int dx\,dy\,\mathbf{W}_l\left[\frac{\partial}{\partial x}\mathbf{D}\frac{\partial}{\partial x}+\frac{\partial}{\partial y}\mathbf{D}\frac{\partial}{\partial y}-\mathbf{R}_a-\mathbf{R}_s+\mathbf{S}\right]\boldsymbol{\varphi}_n, \tag{1.32}$$

$$\mathbf{F}_{ln}^T(z,t)\equiv\int dx\,dy\,\mathbf{W}_l\boldsymbol{\chi}_P\mathbf{F}^T\boldsymbol{\varphi}_n, \tag{1.33}$$

$$\mathbf{F}_{m,ln}^T(z,t)\equiv\int dx\,dy\,\mathbf{W}_l\boldsymbol{\chi}_m\mathbf{F}^T\boldsymbol{\varphi}_n, \tag{1.34}$$

$$\boldsymbol{\tau}_{ln}(z)\equiv\int dx\,dy\,\mathbf{W}_l\mathbf{V}^{-1}\boldsymbol{\varphi}_n, \tag{1.35}$$

$$\mathbf{Q}_l(z,t)\equiv\int dx\,dy\,\mathbf{W}_l\mathbf{Q}, \tag{1.36}$$

$$\mathbf{C}_{m,l}(z,t)\equiv\int dx\,dy\,\mathbf{W}_l\boldsymbol{\chi}_m C_m. \tag{1.37}$$

Equations (1.29) are coupled partial differential equations and Equations (1.30) are coupled ordinary differential equations. These equations are solved for $\mathbf{a}_{ln}$ and $\mathbf{C}_{m,l}$, and the neutron flux is constructed from a relation similar to relation (1.19). Physically, Equations (1.29) and (1.30) require that the modal expansion satisfy Equations (1.3) and (1.4) in N weighted integral senses, with the integral taken over x and y. The finite-difference approximation is usually applied to the z-dependence to reduce Equations (1.29) to coupled ordinary differential equations in time which may be solved by one of the numerical integration methods discussed in Chapter 2.

A further approximation, referred to as "group collapsing," is sometimes employed in conjunction with the modal expansion method. In this approximation, two or more group fluxes of a particular expansion

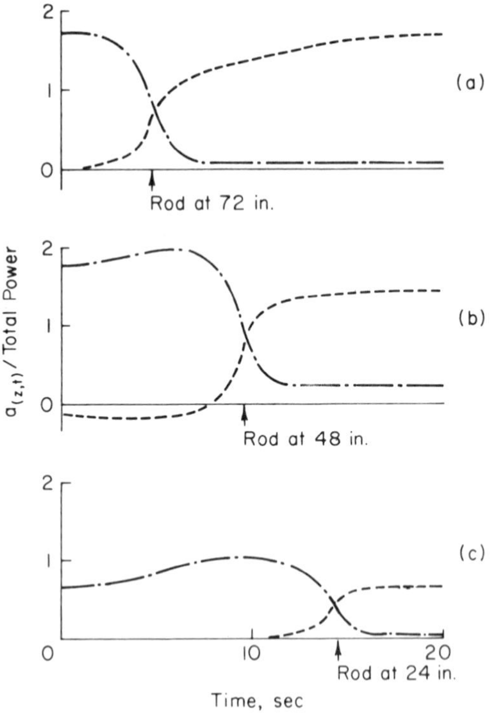

FIGURE 1.1. Modal expansion coefficients for the space-time synthesis of a rod drive-down transient. (*a*) 72-in. elevation; (*b*) 48-in. elevation, at mode discontinuity; (*c*) 24-in. elevation. KEY: - - -, rodded mode; – - –, unrodded mode.

function are required to have the same expansion coefficient; i.e., $a_n^g = a_n^{g'}$ $(g' \neq g)$. The case in which all the group fluxes of an expansion function have the same expansion coefficient (i.e., $a_n^1 = a_n^2 = \cdots = a_n^G$) is known as total group collapsing. Group collapsing reduces the number of unknown expansion coefficients, at the same time reducing the flexibility of the approximation. As the number of unknowns is reduced, it is necessary to reduce the number of equations. This reduction is generally accomplished by summing the individual group synthesis equations corresponding to a given weighting function ($\mathbf{W}_l$) over the collapsed groups. In this case, the relative magnitudes of the different group components in the weighting function determines the relative importance of the individual group balance equations being satisfied in a weighted integral sense, since the physical significance of the collapsed group synthesis equations is that the weighted sum of the group balance equations is satisfied in an integral sense. Group collapsing has led to some interesting anomalies.[16,17]

Figure 1.1 illustrates the results of a typical space-time synthesis calculation. An 8-ft-high cylindrical reactor model, with annular control rods, was used. Material composition varied radially and was different in the top and bottom halves of the reactor. The appropriate unrodded and rodded static radial flux shapes were used as expansion functions in each half,†

$$\phi^g(r, z, t) = a^g_{\text{unr}}(z, t)\varphi^g_{\text{unr}}(r) + a^g_{\text{rod}}(z, t)\varphi^g_{\text{rod}}(r).$$

Three energy groups were used, and the groups were collapsed (i.e., $a^1_{\text{unr}} = a^2_{\text{unr}} = a^3_{\text{unr}}$ and $a^1_{\text{rod}} = a^2_{\text{rod}} = a^3_{\text{rod}}$).

The transient was simulated by driving the control rod down into an initially unrodded core. Figure 1.1 depicts the expansion coefficients at three axial positions as a function of time. These coefficients were divided by the total power to maintain approximately the same normalization throughout the transient. At each axial elevation, the synthesized radial flux shape corresponds to the unrodded expansion function at times well

† When different sets of expansion functions are used in different axial zones, it is necessary to treat the interface between these zones in a special manner. Equations (1.29) are valid within the zones, but not at the interface. Interface conditions may be obtained by requiring flux and current continuity in a weighted integral sense over the interfaces. Alternately, variational methods may be used which allow different sets of expansion functions in different axial zones. This latter is discussed in Chapter 3.

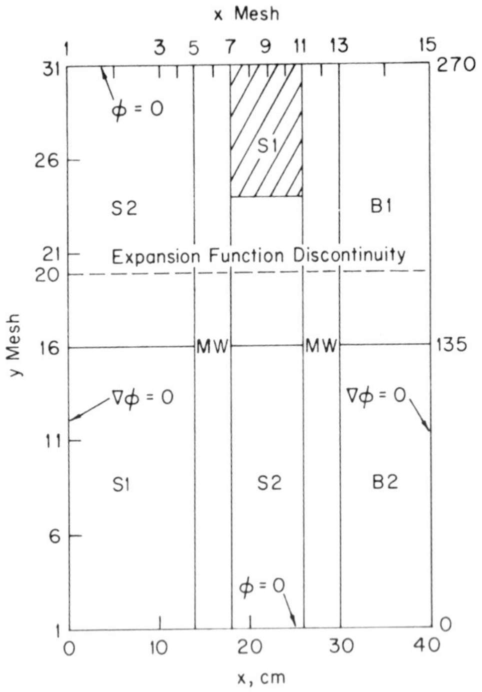

FIGURE 1.2. Reactor model for space-time synthesis calculation.

before the rod reaches that elevation. As the rod approaches a particular elevation, the radial flux shape gradually shifts from the unrodded to the rodded expansion function. At times well after the rod has reached a given elevation, the radial flux shape corresponds almost entirely to the rodded expansion function.

Results of another space-time synthesis calculation, performed on the two-group two-dimensional (x-y) reactor model illustrated in Figure 1.2, are compared with the numerical solution of the time-dependent finite difference equations in Figure 1.3.† Regions S1 and S2 have high multiplication properties, regions B1 and B2 have low multiplication properties,

† The space-time synthesis solution is based on the time-integrated numerical integration algorithm of Section 2.4, and the finite-difference solution is based on the θ-algorithm of Section 2.3.

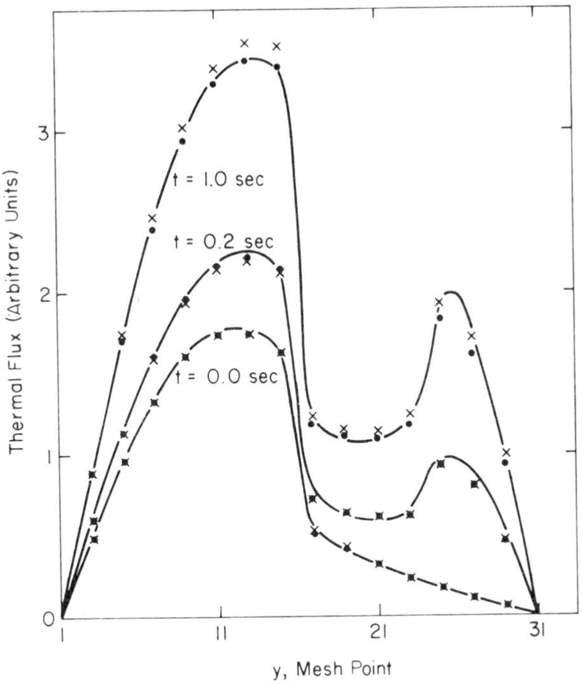

FIGURE 1.3. Comparison of space-time synthesis (STINT) and finite-difference (TWIGL) calculations. KEY: —, TWIGL; ●, STINT, continuous expansion functions; ×, STINT, discontinuous expansion functions.

and MW refers to a metal and water region. The transient was induced by changing the absorption and scattering properties of the cross-hatched region over the period 0.0–0.2 sec. The resulting transient was in the delayed critical range.

Three expansion functions were employed in the space-time synthesis model (for the case labeled "continuous synthesis"),

$$\phi^g(x, y, t) = \sum_{n=1}^{3} a_n{}^g(y, t)\varphi_n{}^g(x).$$

$\varphi_1{}^g$ and $\varphi_2{}^g$ were one-dimensional (x) static flux shapes corresponding to the composition in the lower and upper halves of the model, respectively, and $\varphi_3{}^g$ was a one-dimensional static flux shape corresponding to the upper part of the reactor with the perturbation in the cross-hatched region. In the case labeled "discontinuous synthesis," the flux was synthesized from φ_1^g and φ_2^g in the lower half, and from φ_2^g and φ_3^g in the upper half.

The expansion functions were used as weighting functions in both examples.

The choice of modes (expansion functions) and weighting functions is an important aspect of the application of modal expansion methods. Several types of modes that have been used will be described. First, some operators are defined:

$$\mathbf{L} \equiv -\nabla \cdot \mathbf{D}\, \nabla + \mathbf{R}_a + \mathbf{R}_s - \mathbf{S}, \tag{1.38}$$

$$\mathbf{M}_P \equiv \boldsymbol{\chi}_P \mathbf{F}^T. \tag{1.39}$$

Natural Modes The eigenfunctions of the time-dependent neutron and precursor balance operator are referred to as the natural modes, which satisfy

$$\begin{bmatrix} \{(1-\beta)\mathbf{M}_P - \mathbf{L}\} & \lambda_1 \boldsymbol{\chi}_1 & \cdots & \lambda_M \boldsymbol{\chi}_M \\ \beta_1 \mathbf{F}^T & -\lambda_1 & \cdots & 0 \\ \vdots & & \ddots & \\ \beta_M \mathbf{F}^T & 0 & \cdots & -\lambda_M \end{bmatrix} \begin{bmatrix} \boldsymbol{\Psi}_n \\ C_{1,n} \\ \vdots \\ C_{M,n} \end{bmatrix} = \begin{bmatrix} \mathbf{V}^{-1} & 0 & \cdots & 0 \\ 0 & 1 & \cdots & 0 \\ \vdots & & \ddots & \vdots \\ 0 & & \cdots & 1 \end{bmatrix} \omega_n \begin{bmatrix} \boldsymbol{\Psi}_n \\ C_{1,n} \\ \vdots \\ C_{M,n} \end{bmatrix}, \tag{1.40}$$

where $\boldsymbol{\Psi}_n$ is a column vector of group eigenfluxes. Equation (1.40) may be written

$$\mathbf{H}\boldsymbol{\varphi}_n = \omega_n \mathbf{T} \boldsymbol{\varphi}_n. \tag{1.41}$$

The natural modes satisfy a useful biorthogonality property

$$\langle \boldsymbol{\varphi}_k^* \,|\, \mathbf{T} \,|\, \boldsymbol{\varphi}_n \rangle = 0, \qquad k \neq n, \tag{1.42}$$

where $\boldsymbol{\varphi}_k^*$ satisfies

$$\mathbf{H}^* \boldsymbol{\varphi}_k^* = \omega_k \mathbf{T} \boldsymbol{\varphi}_k^*, \tag{1.43}$$

and $\mathbf{H}^*$ is the Hermitian adjoint of $\mathbf{H}$.

Note that Equations (1.3) and (1.4) may be written

$$\mathbf{H}\boldsymbol{\Psi} + \mathbf{Q}' = \mathbf{T}\dot{\boldsymbol{\Psi}}, \tag{1.44}$$

and, because of the orthogonality property, the time-synthesis equations are decoupled,

$$\langle \boldsymbol{\varphi}_n^* \,|\, \mathbf{T} \,|\, \boldsymbol{\varphi}_n \rangle \omega_n \mathbf{a}_n'(t) + \langle \boldsymbol{\varphi}_n^* \,|\, \mathbf{Q}'(t) \rangle = \langle \boldsymbol{\varphi}_n^* \,|\, \mathbf{T} \,|\, \boldsymbol{\varphi}_n \rangle \dot{\mathbf{a}}_n'(t), \tag{1.45}$$

when $\boldsymbol{\varphi}_n^*$ is taken as the weighting function. Here $\mathbf{a}_n'$ is a column vector of the $\mathbf{a}_n(t)$ and the $\mathbf{C}_{m,n}$ of Equations (1.21) and (1.22). The significant feature of Equation (1.45) is the absence of modal coupling. The equation for each modal expansion coefficient (i.e., the set $\{\mathbf{a}_n, \mathbf{C}_{m,n}\}$) is independent of modes with other n' indexes. This property is referred to as finality.[18]

Equation (1.44) actually pertains only to the special case in which the nuclear properties are constant and equal to the nuclear properties used in calculating the natural modes. When the nuclear properties change in time, or when they are different from the properties used in calculating the natural modes, then the special form of Equation (1.45) no longer obtains and the general form of Equations (1.21) and (1.22) with coupling among modes must be used.

The natural modes for problems with temperature and xenon feedback can be generated in an analogous fashion, although the problem is complicated by the nonlinearities. An approximation to the natural modes, the inhour modes,[10] are obtained by requiring all functions $\{\Psi_n^g, C_{m,n}\}$ with the same mode index n to have the same spatial shape. Special properties can be demonstrated for the inhour modes.

Omega Modes The eigenfunctions of the time-dependent neutron balance operator (ignoring delayed neutrons)

$$(\mathbf{M}_P - \mathbf{L})\boldsymbol{\Psi}_n = \omega_n \mathbf{V}^{-1}\boldsymbol{\Psi}_n \tag{1.46}$$

are known as the omega modes.[19] These functions are biorthogonal with respect to $\boldsymbol{\Psi}_n^*$ which satisfy

$$(\mathbf{M}_P - \mathbf{L})^*\boldsymbol{\Psi}_n^* = \omega_n \mathbf{V}^{-1}\boldsymbol{\Psi}_n^*. \tag{1.47}$$

The biorthogonality relation

$$\langle \boldsymbol{\Psi}_k^* \,|\, \mathbf{V}^{-1} \,|\, \boldsymbol{\Psi}_n \rangle = 0, \tag{1.48}$$

$$k \neq n,$$

leads to some simplification in Equations (1.21) and (1.22), but does not decouple the different modal equations (i.e., does not lead to finality).

Lambda Modes Lambda modes are eigenfunctions of the static neutron balance operator[19]:

$$\mathbf{L}\boldsymbol{\Psi}_n = \frac{1}{\lambda_n}\mathbf{M}_P\boldsymbol{\Psi}_n, \tag{1.49}$$

and have the biorthogonality property

$$\langle \boldsymbol{\Psi}_k^* \,|\, \mathbf{M}_P \,|\, \boldsymbol{\Psi}_n \rangle = 0, \qquad k \neq n, \tag{1.50}$$

where $\boldsymbol{\Psi}_k^*$ satisfies

$$\mathbf{L}^*\boldsymbol{\Psi}_n^* = \frac{1}{\lambda_n}\mathbf{M}_P^*\boldsymbol{\Psi}_n^*. \tag{1.51}$$

These modes do not have the property of finality for Equations (1.21) and (1.22).

Green's Functions Modes These modes are not eigenfunctions, but are obtained by dividing the reactor into regions and setting the fission cross section to zero, then inserting a source into each region and calculating the resulting flux distribution.[20] The source is usually distributed as the fundamental fission distribution, but other choices have been used. The flux shape calculated by putting a source in a region is referred to as the Green's function mode for that region. The Green's functions modes can also be used as the weighting functions.

Synthesis Modes These modes might more aptly be called intuitive modes, for their choice is enhanced by knowledge of the problem. The idea is to pick a set of shapes from which any instantaneous flux distribution that occurs during the transient may be constructed by linear combination. Usually, extreme flux distributions that might arise because of rod motion or feedback are anticipated and used as synthesis modes. For example, a rod insertion transient at low power might be synthesized from rodded and unrodded static flux shapes. Weighting functions may be static adjoint functions corresponding to the extreme configurations, or the synthesis modes themselves may be used as weighting functions.

Analytical Functions The flux within a given region of a reactor may be expanded in a polynomial which is characterized by the value of the flux at several mesh points within the region, the latter being the unknowns in the resulting synthesis equation.[21,22] Alternately, the flux within a given region may be expanded in eigenfunctions of the Helmholtz equation, which, for regular geometries, are analytical functions.[23]

Modal expansion methods, employing all of the previously mentioned weighting and expansion functions, have been successfully applied to idealized reactor models,[11,15,23–28,39] and the space-time synthesis method with synthesis modes has been successfully applied to realistic (3D, $\sim 10^4$ regions, 3 energy group) reactor models. Some of the advantages and disadvantages of the various methods are discussed in the following paragraphs.

The natural modes are, in general, complex. This has limited their successful application either to very idealized multidimensional models or to one-dimensional models. Attempts to construct two-dimensional natural modes by synthesizing one-dimensional modes have been encouraging, if not altogether successful.[11] These modes are complete and possess the property of finality.

Both the omega and lambda modes involve the standard production and destruction operators of group-diffusion theory, and can, in principle, be obtained with the use of standard computer codes. The fundamental lambda mode is, of course, the flux of the static neutron group-diffusion eigenvalue problem. Higher lambda modes can be obtained if symmetry considerations allow node surfaces to be identified or if the Wielandt fractional iteration scheme is used. The computation of omega modes generally requires some substantial modification of the static group-diffusion code.

Green's functions modes are readily obtained with standard group-diffusion codes. These modes are the flux that results when a source is distributed within a given region, and thus seem capable intuitively of representing the effect upon the overall flux distribution of a localized disturbance. The symmetry properties of a reactor can be utilized to minimize the number of Green's functions modes that must be calculated for a given problem.

Synthesis modes are chosen generally to incorporate into the calculational model any information about the solution that is known. They require a certain amount of clairvoyance, which comes with experience, but enable a great deal of information to be obtained from the solution of a relatively simple set of equations. Suitable synthesis modes are frequently available as the result of other calculations. For detailed models, for which the calculation of a single mode may require a few hours of computer time, this can become a compelling argument for the use of synthesis modes.

The physical property of an everywhere nonnegative fundamental mode flux, which was shown in Section 1.2 to be an intrinsic mathematical property of the multigroup finite-difference kinetics equations, is not, in general, an intrinsic mathematical property of the multigroup synthesis kinetics equations. Cases have been examined in which the flux eigenfunction corresponding to the algebraically largest eigenvalue of the collapsed group synthesis equations was negative in some parts of the reactor model.[16,17] This type of anomaly can usually be traced to a physically unreasonable choice of weighting functions. The bulk of experience with the modal expansion approximations indicates that the method yields accurate results when reasonable expansion functions and weighting functions are used.

1.4 Nodal Approximation

The two previous approximations (finite-difference and modal) were oriented toward obtaining a great deal of information about the transient power distribution, and are, in principle, capable of yielding the time-dependent power density at a large number of points in the reactor. Solution of the finite-difference equations presents a formidable task for realistic reactor models, and the calculation of modes and weighting functions and evaluation of the necessary integrals can also be rather time consuming.

A less ambitious, and, consequently, simpler, approximation results when the reactor is visualized as consisting of a relatively small number of coupled regions, and the calculation is oriented toward obtaining the average flux or power level in each region. This visualization is quite appropriate for reactor systems which consist of an array of almost critical cores dispersed in a medium with much inferior multiplicative properties.

Nodal kinetics equations have been derived in a variety of ways, ranging from purely physical[29] to purely mathematical[30] constructions. A coherent review of these derivations was recently published.[31] In this section, a general formalism will be developed which contains the essential ideas of the nodal approximation and which emphasizes the similarities between the nodal and modal approximations. The starting point for the derivation is Equations (1.3) and (1.4). The group flux within each node R_j is written as the product of a shape function $\Psi_j^g(x, y, z)$ and an amplitude function $N_j^g(t)$,

$$\phi(x, y, z, t) = \begin{bmatrix} \Psi_j^1(x, y, z) & 0 & \cdots & 0 \\ 0 & \Psi_j^2(x, y, z) & \cdots & 0 \\ \vdots & & \ddots & \vdots \\ 0 & \cdots & & \Psi_j^G(x, y, z) \end{bmatrix} \begin{bmatrix} N_j^1(t) \\ \vdots \\ N_j^G(t) \end{bmatrix}$$

or

$$\phi(x, y, z, t) = \boldsymbol{\Psi}_j(x, y, z)\mathbf{N}_j(t), \qquad xyz \in R_j. \tag{1.52}$$

Equations for node j are obtained by weighting Equation (1.3) with $\mathbf{W}_j$ and Equations (1.4) by $\mathbf{W}_j\chi_m$, where

$$\mathbf{W}_j(x, y, z) = \begin{bmatrix} W_j^1(x, y, z) & 0 & \cdots & 0 \\ 0 & W_j^2(x, y, z) & \cdots & 0 \\ \vdots & & \ddots & \vdots \\ 0 & \cdots & & W_j^G(x, y, z) \end{bmatrix} \tag{1.53}$$

is an arbitrary weighting function, and integrating over the volume of node $j(R_j)$,

$$\tau_j\dot{\mathbf{N}}_j(t) = \mathbf{L}_j(t) - \mathbf{A}_j(t)\mathbf{N}_j(t) + (1-\beta)\mathbf{F}_j^T(t)\mathbf{N}_j(t) + \sum_{m=1}^{M} \lambda_m \mathbf{C}_{m,j}(t) + \mathbf{Q}_j(t), \qquad j = 1, \ldots \text{ nodes}, \tag{1.54}$$

$$\beta_m \mathbf{F}_{m,j}^T(t)\mathbf{N}_j(t) - \lambda_m \mathbf{C}_{m,j}(t) = \dot{\mathbf{C}}_{m,j}(t), \tag{1.55}$$

$$m = 1, \ldots, M; \qquad j = 1, \ldots \text{ nodes},$$

where

$$\tau_j \equiv \int_{R_j} dx\, dy\, dz\, \mathbf{W}_j \mathbf{V}^{-1} \boldsymbol{\Psi}_j, \tag{1.56}$$

$$\mathbf{A}_j \equiv \int_{R_j} dx\, dy\, dz\, \mathbf{W}_j \{\mathbf{R}_a + \mathbf{R}_s - \mathbf{S}\} \boldsymbol{\Psi}_j, \tag{1.57}$$

$$\mathbf{F}_j{}^T \equiv \int_{R_j} dx\,dy\,dz\,\mathbf{W}_j \chi_P \mathbf{F}^T \boldsymbol{\Psi}_j, \tag{1.58}$$

$$\mathbf{F}^T_{m,j} \equiv \int_{R_j} dx\,dy\,dz\,\mathbf{W}_j \chi_m \mathbf{F}^T \boldsymbol{\Psi}_j, \tag{1.59}$$

$$\mathbf{Q}_j \equiv \int_{R_j} dx\,dy\,dz\,\mathbf{W}_j \mathbf{Q}, \tag{1.60}$$

$$\mathbf{C}_{m,j} \equiv \int_{R_j} dx\,dy\,dz\,\mathbf{W}_j \chi_m C_m, \tag{1.61}$$

$$\mathbf{L}_j \equiv \int_{S_j} d\hat{s} \cdot \mathbf{W}_j \mathbf{D}\,\nabla\phi - \int_{R_j} dx\,dy\,dz\,\nabla\mathbf{W}_j \cdot \mathbf{D}\,\nabla\boldsymbol{\Psi}_j. \tag{1.62}$$

The quantity $\mathbf{L}_j$ results from an application of Gauss's theorem to the diffusion term. The first term in (1.62) defines an integral over the surface S_j of node j of the outward normal component of the weighted current, $\mathbf{W}_j\mathbf{D}\nabla\phi$. Evaluation of $\mathbf{L}_j$ requires an assumption of the net neutron current $(-\mathbf{D}\nabla\phi)$ on the bounding surface of the node, as well as an assumption about the flux shape within the node. The $\mathbf{L}_j$ term represents the contribution of neutrons that diffuse in and out of node j to the neutron balance in node j. For computational purposes, $\mathbf{L}_j$ is approximated by a relation of the form

$$\mathbf{L}_j(t) \cong \sum_{i \neq j} \boldsymbol{l}_{ij}(t)[\mathbf{N}_i(t) - \mathbf{N}_j(t)] \tag{1.63}$$

in which the sum is over all other nodes except j, and $\boldsymbol{l}_{ij}$ is a diagonal matrix whose elements l^g_{ij} are "coupling coefficients" for neutrons in group g between nodes i and j.

Group collapsing can also be used with the nodal approximation. Total group collapsing corresponds to setting $N_j{}^1 = N_j{}^2 = \cdots = N_j{}^G$ for each j and adding the G equations represented by each of Equations (1.54) and (1.55), which yields

$$\begin{aligned} \frac{1}{v_j}\dot{N}_j(t) = \sum_{i \neq j} l_{ij}(t)[N_i(t) - N_j(t)] + \frac{\rho_j(t) - \beta}{v_j \Lambda_j(t)} N_j(t) \\ + \sum_{m=1}^{M} \lambda_m C_{m,j}(t) + Q_j(t), \end{aligned} \tag{1.64}$$

$$\begin{aligned} \dot{C}_{m,j}(t) = \frac{\beta_m}{v_j \Lambda_{m,j}(t)} N_j(t) - \lambda_m C_{m,j}(t), \\ m = 1, \ldots, M, \end{aligned} \tag{1.65}$$

where

$$\frac{1}{v_j} \equiv \int_{R_j} \mathbf{W}_j{}^T \mathbf{V}^{-1} \boldsymbol{\Psi}_j \, dx\, dy\, dz \tag{1.66}$$

$$l_{ij} \equiv \sum_{g=1}^{G} l_{ij}^g \tag{1.67}$$

$$\rho_j \equiv \frac{\int_{R_j} \mathbf{W}_j{}^T \{\chi_P \mathbf{F}^T - \mathbf{R}_a - \mathbf{R}_s + \mathbf{S}\} \boldsymbol{\Psi}_j \, dx\, dy\, dz}{\int_{R_j} \mathbf{W}_j{}^T \chi_P \mathbf{F}^T \boldsymbol{\Psi}_j \, dx\, dy\, dz} \tag{1.68}$$

$$\Lambda_j \equiv \frac{1}{v_j \int_{R_j} \mathbf{W}_j{}^T \chi_P \mathbf{F}^T \boldsymbol{\Psi}_j \, dx\, dy\, dz} \tag{1.69}$$

$$\Lambda_{m,j} \equiv \frac{1}{v_j \int_{R_j} \mathbf{W}_j{}^T \chi_m \mathbf{F}^T \boldsymbol{\Psi}_j \, dx\, dy\, dz} \tag{1.70}$$

$$Q_j \equiv \int_{R_j} \mathbf{W}_j{}^T \mathbf{Q} \, dx\, dy\, dz \tag{1.71}$$

$$C_{m,j} \equiv \int_{R_j} \mathbf{W}_j{}^T \chi_m C_m \, dx\, dy\, dz. \tag{1.72}$$

In Equations (1.66)–(1.72), $\mathbf{W}_j$ and $\boldsymbol{\Psi}_j$ are $G \times 1$ column vectors with elements W_j^g and Ψ_j^g.

Although the parameters of the nodal model can be defined from Equations (1.56)–(1.62), or Equations (1.66)–(1.72), and the coupling coefficients l_{ji}^g can be defined by comparing Equations (1.62) and (1.63), parameters such as v_j, ρ_j, l_{ij}, and Λ_j are often specified from physical considerations without recourse to either the defining relations or the $\mathbf{W}_j$ and $\boldsymbol{\Psi}_j$. The quantity v_j is an average speed, whereas ρ_j is a "nodal reactivity", Λ_j is a nodal prompt neutron generation time and $\Lambda_{m,j}$ is a modified prompt neutron generation time.

The principal difficulty encountered in using the nodal model is associated with the coupling coefficients, l_{ij}^g or l_{ij}. From Equations (1.62) and (1.63), and from physical considerations, it is apparent that these coefficients, which influence the rate at which neutrons diffuse in and out of node j, depend upon the assumed flux shape within node j and on the assumed neutron current on the bounding surface S_j. These coefficients can be evaluated by choosing a flux shape Ψ_j and a weighting function $\mathbf{W}_j$, assuming a surface current, $-\mathbf{D}\,\nabla\phi$, and evaluating the appropriate integrals. Alternately, the l_{ij} may be chosen so that the nodal model correctly predicts a known reaction rate distribution among nodes. In either case, the internodal coupling coefficients are based on a given assumed flux shape. When the reactor configuration changes, coupling coefficients based upon an unperturbed flux shape are no longer appropriate, and have been found to introduce a significant error into the calculation of transient flux tilting.

Although it is possible, in principle, to change the coupling coefficients during a transient, the manner in which they should be changed is not generally known. This difficulty has been at least partially circumvented by choosing the coupling coefficients so that the nodal calculation matches, as well as possible, several known extremum reaction rate distributions that bound any distributions that might obtain during a transient, rather than to match any single distribution exactly.[32]

The similarity between the nodal and modal approximations is apparent from the formalism. The nodal approximation is actually a special case of a modal approximation in which only a single expansion mode and a single weighting function are used in a given region. The use of a different expansion mode in adjacent nodes, or the use of a continuous expansion mode with different expansion coefficients in adjacent nodes, introduces the possibility of discontinuities in the detailed flux and current distributions constructed from Equation (1.52) at node interfaces. Since the nodal model is normally employed only to obtain the average flux level in a node ($\propto N_j$), this is not a practical problem. The similarity between nodal and modal models and the problem of discontinuities are discussed further in Chapter 3.

1.5 Point Kinetics

An even less ambitious, but very popular, approximation results when the calculation is oriented toward obtaining only the total power or neutron flux level in the reactor. Although in general this method is employed with an implicit assumption of a fixed spatial distribution, this assumption is not essential. The point kinetics equations have been derived many times, and their assumptions and limitations have been discussed in great detail.[33–35] In this section, a derivation is given which underscores the relation to the previous approximations. Several methods for incorporating space-time effects into the point kinetics model are discussed.

Equations (1.3) and (1.4) are the starting point of this derivation. The group flux is written as the product of a shape function $\Psi^g(x, y, z, t)$, which may be time dependent, and an amplitude function $N^g(t)$:

$$\boldsymbol{\phi}(x, y, z, t) \equiv \begin{bmatrix} \Psi^1(x, y, z, t) & 0 & \cdots & 0 \\ 0 & \Psi^2(x, y, z, t) & \cdots & 0 \\ \vdots & & \ddots & \vdots \\ 0 & & \cdots & \Psi^G(x, y, z, t) \end{bmatrix} \begin{bmatrix} N^1(t) \\ \vdots \\ N^G(t) \end{bmatrix}. \tag{1.73}$$

No approximation is involved in expressing the flux in this manner.

The general point kinetics equations are obtained by first defining group weighting functions $W^g(x, y, z, t)$,

$$\mathbf{W}(x, y, z, t) = \begin{bmatrix} W^1(x, y, z, t) & 0 \cdots & 0 \\ \vdots & \ddots & \vdots \\ 0 & \cdots & W^G(x, y, z, t) \end{bmatrix}. \tag{1.74}$$

Then Equation (1.73) is substituted into Equations (1.3) and (1.4), the former is premultiplied by $\mathbf{W}$ and the latter by $\mathbf{W}\chi_m$, and the resulting equations are integrated over the entire volume of the reactor, which yields

$$\begin{aligned} \boldsymbol{\tau}(t)\dot{\mathbf{N}}(t) = & -\mathbf{A}(t)\mathbf{N}(t) + (1 - \beta)\hat{\mathbf{F}}^T(t)\mathbf{N}(t) + \sum_{m=1}^{M} \lambda_m \mathbf{C}_m(t) \\ & + \mathbf{Q}(t) - \mathbf{E}(t)\mathbf{N}(t), \end{aligned} \tag{1.75}$$

$$\dot{\mathbf{C}}_m(t) = \beta_m \hat{\mathbf{F}}_m{}^T \mathbf{N}(t) - \lambda_m \mathbf{C}_m(t), \tag{1.76}$$

$$m = 1, \ldots, M,$$

where

$$\boldsymbol{\tau}(t) \equiv \int dx\, dy\, dz\, \mathbf{W}\mathbf{V}^{-1}\boldsymbol{\Psi}, \tag{1.77}$$

$$\mathbf{A}(t) \equiv \int dx\, dy\, dz\, \mathbf{W}\{-\nabla \cdot \mathbf{D}\,\nabla + \mathbf{R}_a + \mathbf{R}_s - \mathbf{S}\}\boldsymbol{\Psi}, \tag{1.78}$$

$$\hat{\mathbf{F}}^T(t) \equiv \int dx\, dy\, dz\, \mathbf{W}\chi_p \mathbf{F}^T\boldsymbol{\Psi}, \tag{1.79}$$

$$\hat{\mathbf{F}}_m{}^T(t) \equiv \int dx\, dy\, dz\, \mathbf{W}\chi_m \mathbf{F}^T\boldsymbol{\Psi}, \tag{1.80}$$

$$\hat{\mathbf{Q}}(t) \equiv \int dx\, dy\, dz\, \mathbf{W}\mathbf{Q}, \tag{1.81}$$

$$\mathbf{C}_m(t) \equiv \int dx\, dy\, dz\, \mathbf{W} \boldsymbol{\chi}_m C_m, \tag{1.82}$$

$$\mathbf{E}(t) \equiv \int dx\, dy\, dz\, \mathbf{W} \mathbf{V}^{-1} \dot{\boldsymbol{\Psi}}. \tag{1.83}$$

The time dependence of the quantities defined by Equations (1.77)–(1.82) is in part due to the time dependence of the nuclear properties, source, and precursor density, and in part due to the time dependence of the weighting function and the shape function. In the conventional point kinetics formalism, $\mathbf{E} \equiv 0$. This may be accomplished by requiring that the shape function is constant in time, in which case Equations (1.73) is an approximate, rather than exact, relation. Alternately, $\boldsymbol{\Psi}$ may be normalized such that the expression in Equation (1.83) vanishes.[35] In the usual case when the weighting function is time independent, the condition that $\mathbf{E} = 0$ imposes the following normalization on $\boldsymbol{\Psi}$:

$$\mathbf{E}(t) = \frac{\partial}{\partial t} \int dx\, dy\, dz\, \mathbf{W} \mathbf{V}^{-1} \boldsymbol{\Psi} = 0. \tag{1.84}$$

Group collapsing can be used with the point kinetics approximation. For total group collapsing, the G Equations (1.84) are added, as are the G Equations (1.85) for each value of m:

$$\frac{1}{v(t)} \dot{N}(t) = \frac{\rho(t) - \beta}{v(t)\Lambda(t)} N(t) + \sum_{m=1}^{M} \lambda_m \hat{C}_m(t) + \hat{Q}(t), \tag{1.85}$$

$$\dot{\hat{C}}_m(t) = \frac{\beta_m}{v(t)\Lambda_m(t)} N(t) - \lambda_m \hat{C}_m(t), \qquad m = 1, \ldots, M, \tag{1.86}$$

where

$$\frac{1}{v(t)} \equiv \int dx\, dy\, dz\, \mathbf{W}^T \mathbf{V}^{-1} \boldsymbol{\Psi}, \tag{1.87}$$

$$\Lambda(t) \equiv \frac{1}{v(t) \int dx\, dy\, dz\, \mathbf{W}^T \boldsymbol{\chi}_p \mathbf{F}^T \boldsymbol{\Psi}}, \tag{1.88}$$

$$\Lambda_m(t) \equiv \frac{1}{v(t) \int dx\, dy\, dz\, \mathbf{W}^T \boldsymbol{\chi}_m \mathbf{F}^T \boldsymbol{\Psi}}, \tag{1.89}$$

$$\rho(t) \equiv \frac{\int dx\, dy\, dz\, \mathbf{W}^T\{\chi_P \mathbf{F}^T + \nabla \cdot \mathbf{D}\, \nabla - \mathbf{R}_a - \mathbf{R}_s + \mathbf{S}\}\boldsymbol{\Psi}}{\int dx\, dy\, dz\, \mathbf{W}^T \chi_P \mathbf{F}^T \boldsymbol{\Psi}}, \tag{1.90}$$

$$\hat{Q}(t) \equiv \int dx\, dy\, dz\, \mathbf{W}^T \hat{\mathbf{Q}}, \tag{1.91}$$

$$\hat{C}_m(t) \equiv \int dx\, dy\, dz\, \mathbf{W}^T \chi_m C_m, \tag{1.92}$$

and the normalization condition

$$E(t) \equiv \int dx\, dy\, dz\, \mathbf{W}^T \mathbf{V}^{-1} \dot{\boldsymbol{\Psi}} = 0 \tag{1.93}$$

is required. In the definitions of Equations (1.87)–(1.93), $\mathbf{W}$ and $\boldsymbol{\Psi}$ are, respectively, $G \times 1$ column vectors of group weighting functions and shape functions.

The quantity ρ is referred to as the reactivity and Λ is known as the prompt neutron generation time. In most derivations, an effective delay fraction is introduced to account for the fact that delayed and prompt neutrons have different spectra. This fission spectra difference is accounted for here in the modified prompt neutron generation time, Λ_m.

Although the weighting function is arbitrary, the static adjoint corresponding to the initial configuration is used traditionally. Choice of the weighting function may be motivated by the consideration that the point kinetics equations essentially are requirements that expressions of the form of Equation (1.73) satisfy Equations (1.3) and (1.4) in a weighted integral sense. In this vein, it should be noted that the quantity obtained as a solution to the point kinetics equations is not the total neutron population or total power, but is the amplitude function.

In the most common application of the point kinetics approximation, both the weighting function and shape function are time independent, and are usually taken as the adjoint and direct eigenfunctions, respectively, of the initial configuration. In this case, the condition $\mathbf{E} \equiv 0$ is satisfied, and the point kinetics equations describe a transient neutron flux that is fixed in spatial distribution but is varying in amplitude.

The adiabatic model[36] improves on this treatment by allowing the shape function to change at selected times during the transient, while requiring the normalization condition of Equation (1.84) to be satisfied. This approximation is implemented by solving a static eigenvalue problem for the flux distribution corresponding to the instantaneous configuration at several times during a transient, and employing this flux distribution to recompute the point kinetics parameters.

Implicit in the adiabatic model is the assumption that the neutron distribution is in equilibrium with the instantaneous thermodynamic and mechanical configuration. For most applications this is a decent assumption with regard to the prompt neutrons, because the prompt neutrons adjust to a new configuration within a few prompt neutron lifetimes ($\sim 10^{-3}$–10^{-5} sec for thermal neutrons). However, in delayed critical and subcritical transients initiated in reactors with a substantial delayed neutron precursor population, the delayed neutrons are not in equilibrium, but tend to remember previous fission source distributions. This delayed neutron effect tends to retard flux tilts. The magnitude of this effect increases as the reactor size increases, or as the reactor becomes more susceptible to flux tilting.

Quasistatic models have been proposed[37] that improve upon the adiabatic model by accounting explicitly for the delayed neutrons having a different spatial shape than the prompt neutrons. The neutron flux is factored into an amplitude function and a shape function, as indicated by Equation (1.73). This factorization is substituted into Equation (1.3) to obtain an equation for the shape function in terms of the amplitude function and the precursor densities. The calculation proceeds by (1) using the initial static shape to evaluate the point kinetics parameters, (2) integrating the point kinetics equations over a time interval to obtain the amplitude functions and percursor densities at time t_1, (3) solving the spatially-dependent equation for the shape function at t_1, (4) re-evaluating the point kinetics parameters using this shape function, and (5) repeating steps 2–5 over many sequential time intervals.

1.6 Discussion

In deciding which of the preceding approximations is appropriate for a given class of problems, the amount of detail that is required in the solution is naturally a leading consideration. Another factor, and one much more difficult to evaluate, is the probable influence of spatial flux shifting upon the solution and the adequacy of the various approximations in accounting correctly for this flux shifting. Situations are conceivable in which very little spatial detail is required in the solution, but in which an adequate representation of spatial flux shifting requires a considerably more elaborate spatial approximation. Finally, the amount of computational time required to obtain the solution must be considered.

In the finite-difference approximation, the spatial detail is obtained by solving for the transient flux at each of a large number of closely spaced mesh points at each of many time steps. The modal expansion method seeks to construct the transient flux distribution by combining predetermined expansion functions at each of many time steps. These expansion functions may actually be finite-difference calculations of the flux at each of many mesh points. Thus, the modal expansion method is capable of representing the transient spatial flux distribution in the same detail as would be obtained with the finite-difference approximation. Since the number of expansion functions used in a modal expansion approximation is generally much less than the number of mesh points in the finite-difference representation of the same model, the number of equations, and thus the amount of computation time required to obtain a solution, is much less for the modal expansion than for the finite-difference approximation. Because the time required to compute the expansion functions frequently constitutes the major component of computation time for the modal expansion method, the advantage of the modal expansion method over the finite-difference method in computational time may be increased considerably if a large number of transients are calculated with the same set of expansion functions.

The nodal approximation provides only the average flux level in each of several large regions of the reactor. Formally, the nodal equations are identical to the finite-difference equations, and the basic difference in the methods is the scale of the spatial regions involved. Thus, much shorter computation times are associated with the nodal than with the finite-difference approximation. Whereas the nodal model represents spatial flux shifting on a gross basis, the modal expansion model represents this shifting on a detailed basis, but the detail is obtained from predetermined expansion functions. Computation times for nodal and modal models should be about the same if the same number of nodes and modes are used in the respective models.

The point kinetics model requires the least computational effort and, in its basic form, provides no spatial detail in the solution other than the flux shape that was assumed initially, and does not account for spatial flux shifts. Elaborations of the point kinetics model, such as the adiabatic

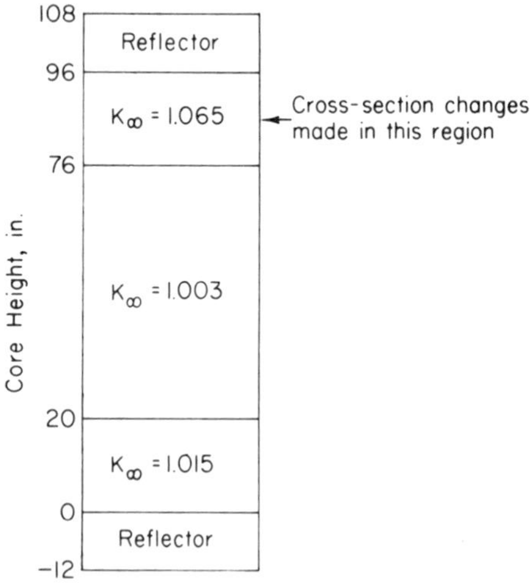

FIGURE 1.4. Reactor model for transient calculations.

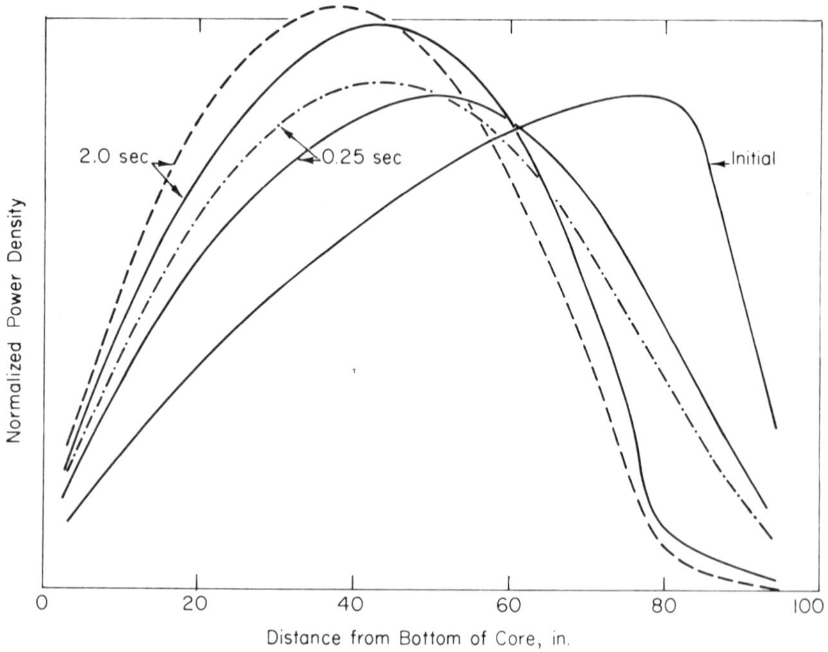

FIGURE 1.5. Power distributions during transient. KEY: —, finite-difference method; – – –, – · – ·, adiabatic (static) calculation.

or quasi-static models, account for spatial flux shifts by recomputing the assumed spatial flux shape at intervals throughout the transient. The amount of spatial detail and computation time associated with these latter models depend upon whether the frequently recomputed spatial flux shapes are based upon the finite-difference, modal expansion, or nodal approximation.

To illustrate how well the more approximate methods fare relative to the more exact finite-difference method when significant spatial flux tilting takes place, several calculations were performed on the one-dimensional, two-group, thermal reactor model shown in Figure 1.4. To initiate a transient in this initially critical model, the absorption cross section was linearly increased in the top core region. Figure 1.5 depicts the power distribution at several times during the transient, and also shows the static power distributions corresponding to the instantaneous material properties at those times. It is apparent that delayed neutrons hold back the power tilt to a significant degree. (Otherwise, the static and dynamic

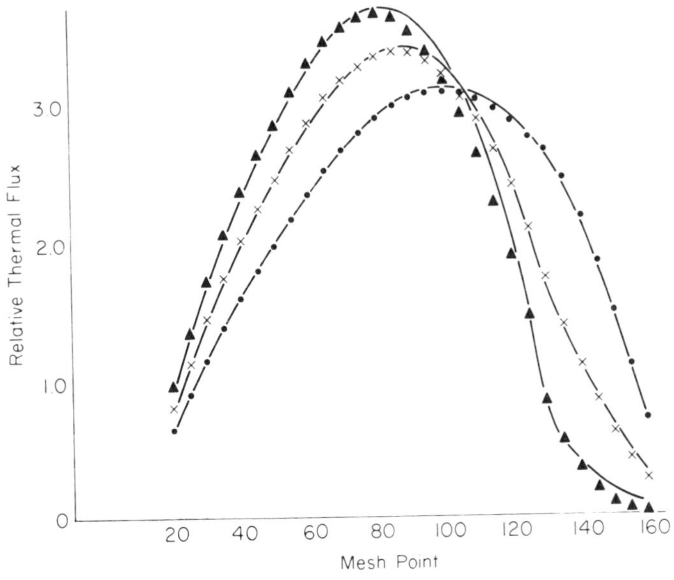

FIGURE 1.6. Thermal flux distributions during transient; 8-ft core, 2-sec ramp. KEY: —, finite difference; synthesis results, adjoint weight, modes 1, 2, 3, 4; ●, 0.1 sec; ×, 0.4 sec; ▲, 1.6 sec.

power shapes would be identical.) These power shapes were based on a finite-difference calculation.

A time-synthesis calculation was performed for this transient. Expansion functions were the initial and final (2.0 sec) static flux shapes shown in Figure 1.5, and static flux shapes calculated with 0.01 and 0.1 of the total increase in absorption cross section which takes place during the transient. The weighting functions were the static adjoint functions calculated for the same four configurations. Thermal flux distributions calculated during the transient compare favorably with finite-difference results as shown in Figure 1.6. These flux distributions are normalized to maintain a constant spatial flux integral; i.e., they do not reflect changes in the total power. Total power predicted by the synthesis model was in good agreement with the finite-difference results.

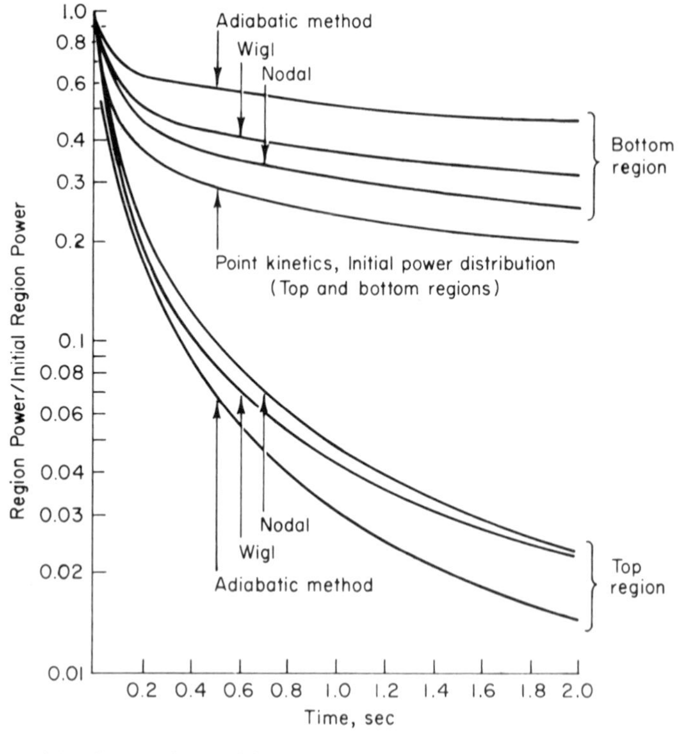

FIGURE 1.7. Comparison of finite-difference (WIGL), nodal, adiabatic, and point kinetics calculations of region powers.

The transient was also calculated with a 12-node model, with a point kinetics model with parameters based upon the initial flux distribution, and with a point kinetics model in which the parameters were periodically recomputed using the adiabatic flux shape (adiabatic model). Figure 1.7 shows the power in the top and bottom core regions as predicted by the three calculations and as predicted by the finite-difference calculation. The adiabatic model is an improvement of the point kinetics model, but the error in the adiabatic flux shapes (see Figure 1.5) owing to delayed neutron holdback of flux tilting causes a corresponding error in the adiabatic calculation. The nodal model is in somewhat better agreement with the finite-difference calculation than is the adiabatic calculation, owing to a better representation of the flux tilting.

Additional comparisons of different methods for space-time computations are given by Ott and Meneley,[37] Yasinsky,[38] and Yasinsky and Henry.[39]

REFERENCES

1. A. M. Weinberg and E. P. Wigner, *The Physical Theory of Neutron Chain Reactors*, Chapter 9. University of Chicago Press, Chicago, 1958.
2. B. Davidson, *Neutron Transport Theory*, Chapters 8 and 19. Oxford University Press (Clarendon), London and New York, 1958.
3. *Reactor Physics Constants*, 2nd ed. ANL-5800, Argonne National Laboratory, 1963.
4. G. J. Habetler and M. A. Martino, "Existence Theorems and Spectral Theory for the Multigroup Diffusion Model," *Proc. Symp. Appl. Math.* **11,** 127–139. Am. Math. Soc., Providence, Rhode Island, 1961.
5. G. R. Keepin, *Physics of Nuclear Kinetics*, Chapters 4 and 5. Addison-Wesley, Reading, Massachusetts, 1965.
6. E. L. Wachspress, *Iterative Solution of Elliptic Systems*. Prentice-Hall, Englewood Cliffs, New Jersey, 1966.
7. R. S. Varga and M. A. Martino, "The Theory for the Numerical Solution of Time-Dependent and Time-Independent Multigroup Diffusion Equations", *Proc. Intern. Conf. Peaceful Uses At. Energy, Geneva* **16,** 570, 1958.
8. R. S. Varga, *Matrix Iterative Analysis*. Prentice-Hall, Englewood Cliffs, New Jersey, 1962.
9. G. Birkhoff and R. A. Varga, "Reactor Criticality and Non-Negative Matrices," WAPD-166, Bettis Atomic Power Laboratory, 1957.
10. A. F. Henry, "The Application of Inhour Modes to the Description of Nonseparable Reactor Transients," *Nucl. Sci. Eng.* **20,** 338 (1964).

11. L. R. Foulke, "A Modal Expansion Technique for Space-Time Reactor Kinetics," Ph.D. thesis, Department of Nuclear Engineering, Massachusetts Institute of Technology (1966); also L. R. Foulke and E. P. Gyftopoulos, "Application of the Natural Model Approximation to Space-Time Reactor Problems," *Nucl. Sci. Eng.* **30,** 419 (1967).
12. T. A. Porsching, "On the Spectrum of a Matrix Arising from a Problem in Reactor Kinetics," *SIAM J. Appl. Math.* **16** (1968).
13. W. M. Stacey, Jr., *Modal Approximations: Theory and an Application to Reactor Physics.* M.I.T. Press, Cambridge, Massachusetts, 1967; also, "A General Modal Expansion Method for Obtaining Approximate Equations for Linear Systems," *Nucl. Sci. Eng.* **28,** 438 (1967).
14. S. Kaplan, "Synthesis Methods in Reactor Analysis," *Advan. Nucl. Sci. Technol.* **3,** 233 (1965).
15. S. Kaplan, O. J. Marlowe and J. Bewick, "Application of Synthesis Techniques to Problems Involving Time Dependence," *Nucl. Sci. Eng.* **18,** 163 (1964).
16. J. B. Yasinsky and S. Kaplan, "Anomalies Arising from the Use of Adjoint Weighting in a Collapsed Group-Space Synthesis Model," *Nucl. Sci. Eng.* **31,** 354 (1968).
17. C. H. Adams and W. M. Stacey, Jr., "An Anomaly Arising in the Collapsed-Group Flux Synthesis Approximation," *Nucl. Sci. Eng.* **36,** 444 (1969).
18. S. Kaplan, "The Property of Finality and the Analysis of Problems in Reactor Space-Time Kinetics by Various Modal Expansions," *Nucl. Sci. Eng.* **9,** 357 (1961).
19. S. Kaplan, "Modal Analysis," Sect. 5.4.D in *Naval Reactors Physics Handbook*, (A. Radkowsky, ed.), Vol. I. TID-7030, USAEC, Washington, D.C., 1964.
20. D. E. Dougherty and C. N. Shen, "The Space-Time Neutron Kinetic Equations Obtained by the Semidirect Variational Method," *Nucl. Sci. Eng.* **13,** 141 (1962).
21. J. W. Riese, "VARI-QUIR: A Two-Dimensional Time Dependent Multigroup Diffusion Code," WANL-TNR-133, Westinghouse Astro-Nuclear Laboratory (1963).
22. K. F. Hansen and S. R. Johnson, "An Approximate Method for Multidimensional Diffusion Theory Problems," GA-7544, General Atomic (1967).
23. R. Bobone, "Analytic Solution of the Time Linearized Kinetic Equations of Multiregion Reactors in the Diffusion Approximation, *Trans. Am. Nucl. Soc.* **10,** 548 (1967).
24. J. A. Bewick and S. Kaplan, "A Test of the Time Synthesis Approach for the Solution of Reactor Kinetics Problems," WAPD-TM-641, Bettis Atomic Power Laboratory (1966).
25. W. G. Clarke and S. G. Margolis, "The Multimode Synthesis Approximation for Space-Time Reactor Dynamics: An Appraisal of Finite Differencing Methods," WAPD-TM-635, Bettis Atomic Power Laboratory (1967).

26. R. A. Rydin, "Time Synthesis—A Study of Synthesis Modes and Weighting Functions," *Trans. Am. Nucl. Soc.* **10,** 569 (1967).
27. R. J. Hooper and M. Becker, "Flux Synthesis Using Modified Green's Function Modes," *Trans. Am. Nucl. Soc.* **9,** 471 (1966).
28. N. Carter and R. Danofsky, "The Application of the Calculus of Variations and The Method of Green's Functions to the Solution of Coupled Core Kinetics Equations," *Proc. Conf. Coupled Reactor Kinetics*, (C. G. Chezem and W. H. Kohler, eds.), p. 249. Texas A & M Press, College Station, Texas, 1967.
29. R. Avery, "Theory of Coupled Reactors," *Proc. Intern. Conf. Peaceful Uses At. Energy, 2nd, Geneva* **12,** 182 (1958).
30. R. G. Cockrell and R. B. Perez, "On the Kinetic Theory of Spatial and Spectral Coupling of the Reactor Neutron Field," *Proc. Symp. Neutron Dynamics and Control.* CONF-650413, USAEC, Washington, D.C., 1966.
31. F. T. Adler, S. J. Gage, and G. C. Hopkins, "Spatial and Spectral Coupling Effects in Multicore Reactor Systems," *Proc. Conf. Coupled Reactor Kinetics*, (C. G. Chezem and W. H. Kohler, eds.). Texas A & M Press, College Station, Texas, 1967.
32. D. C. Wade and H. H. Rubin, "A Nodal Calculation of a Space-Time Transient Using Coupling Coefficients which Account for Changing Internodal Leakages," *Trans. Am. Nucl. Soc.* **10,** 250 (1967).
33. A. F. Henry in *Naval Reactors Physics Handbook* (A. Radkowsky, ed.), Vol. I, Section 5.2. TID-7030, USAEC, Washington, D.C., 1964.
34. E. P. Gyftopoulos in *The Technology of Nuclear Reactor Safety*, (T. J. Thompson and J. G. Beckerly, eds.), Vol. I, Chapter 3. M.I.T. Press, Cambridge, Massachusetts, 1964.
35. M. Becker, "A Generalized Formulation of Point Nuclear Reactor Kinetics Equations," *Nucl. Sci. Eng.* **31,** 458 (1968).
36. A. F. Henry and N. J. Curlee, "Verification of a Method for Treating Neutron Space-Time Problems," *Nucl. Sci. Eng.* **4,** 727 (1958).
37. K. Ott and D. A. Meneley, "Accuracy of the Quasistatic Treatment of Spatial Reactor Kinetics," *Nucl. Sci. Eng.* **36,** 402 (1969).
38. J. B. Yasinsky, "Numerical Studies of Combined Space-Time Synthesis,' *Nucl. Sci. Eng.* **34,** 158 (1968).
39. J. B. Yasinsky and A. F. Henry, "Some Numerical Experiments Concerning Space-Time Kinetics Behavior," *Nucl. Sci. Eng.* **22,** 171 (1965).

Chapter 2

NUMERICAL INTEGRATION METHODS FOR THE TIME DEPENDENCE

In Chapter 1, approximations for the spatial dependence of the multigroup kinetics equations were outlined. This chapter deals with the somewhat less developed area of approximations to the time dependence. Several methods which have been proposed, and in some cases applied successfully, will be discussed. For many of the methods, important questions of numerical stability and truncation error remain unanswered. The first five methods considered are more or less independent of the spatial approximation employed, but the final two methods are explicitly based on the finite-difference approximation. Thus, the basic equations for the first five methods are the source-free versions of Equations (1.1) and (1.2), which may be written as a supermatrix equation

$$\mathbf{H}\mathbf{\Psi} = \dot{\mathbf{\Psi}}. \tag{2.1}$$

2.1 Explicit Method—Forward Difference

The simplest approximation to Equation (2.1) is obtained by a simple forward difference algorithm,

$$\mathbf{\Psi}(p + 1) = \mathbf{\Psi}(p) + \Delta t\, \mathbf{H}(p)\mathbf{\Psi}(p), \tag{2.2}$$

where the argument p denotes the value at time t_p, and $\Delta t = t_{p+1} - t_p$.

In terms of Equations (1.1) and (1.2), this algorithm is

$$\phi^g(p+1) = \phi^g(p) + \Delta t\, v^g\{\nabla \cdot D^g(p)\, \nabla\phi^g(p) - (\Sigma_a{}^g(p) + \Sigma_s{}^g(p))\phi^g(p)$$

$$+ \sum_{g' \neq g}^{G} \Sigma_s^{g'/g}(p)\phi^{g'}(p) + (1-\beta)\chi_P{}^g \sum_{g'=1}^{G} \nu^{g'}\Sigma_f^{g'}(p)\phi^{g'}(p)$$

$$+ \sum_{m=1}^{M} \lambda_m \chi_m{}^g C_m(p)\}, \tag{2.3}$$

$$g = 1, \ldots, G,$$

and for the precursors,

$$C_m(p+1) = C_m(p) + \Delta t\left[\beta_m \sum_{g=1}^{G} \nu^g \Sigma_f{}^g(p)\phi^g(p) - \lambda_m C_m(p)\right], \tag{2.4}$$

$$m = 1, \ldots, M.$$

This algorithm suffers from a problem of numerical stability, which requires the use of such small time steps that the advantage offered by the simplicity of the algorithm is usually more than offset by the large number of time steps required. The nature of this problem is seen by considering an expansion of $\mathbf{\Psi}(p)$ in the eigenfunctions of the operator $\mathbf{H}$ [or its finite-difference analog $\mathbf{K}$ of Equation (1.9)],

$$\mathbf{\Psi}(p) = \sum_n a_n \mathbf{\Omega}_n, \tag{2.5}$$

where

$$\mathbf{H}\mathbf{\Omega}_n = \omega_n \mathbf{\Omega}_n. \tag{2.6}$$

Substituting Equation (2.5) into Equation (2.2) yields

$$\mathbf{\Psi}(p+1) = \sum_n a_n(1 + \omega_n\, \Delta t)\mathbf{\Omega}_n. \tag{2.7}$$

The condition for numerical stability is that the fundamental mode $\mathbf{\Omega}_1$ grow more rapidly than the harmonics $\mathbf{\Omega}_n$, $n \geqq 2$. This requires that

$$|1 + \omega_1\, \Delta t| > |1 + \omega_n\, \Delta t|, \tag{2.8}$$

$$n \geqq 2.$$

In order to ensure this, $|\omega_n\, \Delta t|$ must be much less than unity. The magnitude of the fundamental eigenvalue is on the order of the precursor decay constant, except for highly supercritical transients, in which case small time steps are used. Numerical studies[1] have shown that the smallest eigenvalues can be on the order of $-(v^g\Sigma_a{}^g)$, which can be $\sim -10^4$ for thermal neutrons and $\sim -10^8$ for fast neutrons. Thus, $\Delta t < 10^{-8}$ may be required. When the time derivative terms for the epithermal groups are assumed to vanish (a useful approximation since $1/v^G \gg 1/v^g$, $g \neq G$), $\Delta t < 10^{-4}$ may be required.

Higher order explicit methods (e.g., Runge–Kutta or Adams–Bashford) will probably multiply the computer time required to perform a time step by a factor equal to the order of the approximation while obtaining an improvement in step size which is smaller than the order of the approximation (Hansen,[2] p. 11).

2.2 Implicit Integration—Backwards Difference

The numerical stability problem associated with the previous method can all but be eliminated by the backwards-difference algorithm

$$\mathbf{\Psi}(p+1) = [\mathbf{I} - \Delta t\, \mathbf{H}(p+1)]^{-1}\mathbf{\Psi}(p). \tag{2.9}$$

In terms of Equations (1.1) and (1.2), this algorithm is

$$C_m(p+1) = \frac{C_m(p)}{1+\lambda_m \Delta t} + \frac{\beta_m \Delta t}{1+\lambda_m \Delta t} \sum_{g=1}^{G} \nu^g \Sigma_f{}^g(p+1)\phi^g(p+1), \quad m = 1, \ldots, M, \tag{2.10}$$

$$\begin{aligned}
&\nabla \cdot D^g(p+1)\,\nabla\phi^g(p+1) - (\Sigma_a{}^g(p+1) + \Sigma_s{}^g(p+1))\phi^g(p+1) \\
&\quad + \sum_{g' \neq g}^{G} \Sigma_s^{g'/g}(p+1)\phi^{g'}(p+1) + (1-\beta)\chi_P{}^g \sum_{g'=1}^{G} \nu^{g'}\Sigma_f^{g'}(p+1)\phi^{g'}(p+1) \\
&\quad + \sum_{m=1}^{M} \frac{\lambda_m \chi_m{}^g \beta_m \Delta t}{1+\lambda_m \Delta t} \sum_{g'=1}^{G} \nu^{g'}\Sigma_f^{g'}(p+1)\phi^{g'}(p+1) - \frac{1}{v^g \Delta t}\phi^g(p+1) \\
&= -\frac{1}{v^g \Delta t}\phi^g(p) - \sum_{m=1}^{M} \frac{\lambda_m \chi_m{}^g C_m(p)}{1+\lambda_m \Delta t}, \qquad g = 1, \ldots, G.
\end{aligned} \tag{2.11}$$

An expansion of the type of Equation (2.5) substituted into Equation (2.9) yields

$$\mathbf{\Psi}(p+1) = \sum_n a_n(1 - \Delta t\, \omega_n)^{-1}\mathbf{\Omega}_n \tag{2.12}$$

and the condition for stability is

$$|(1 - \Delta t\, \omega_1)^{-1}| > |(1 - \Delta t\, \omega_n)^{-1}|, \qquad n \geqq 2. \tag{2.13}$$

The method is unconditionally stable if $0 > \mathrm{Re}(\omega_1) > \mathrm{Re}(\omega_n)$, $n \geqq 2$. For $\mathrm{Re}(\omega_1) > 0$, the stability requirement is determined by the requirement that $\boldsymbol{\Psi}(p+1)$ be a positive vector, which necessitates

$$\Delta t < \frac{1}{\omega_1}. \tag{2.14}$$

This requirement is restrictive only for large ω_1 that correspond to fast transients where small time steps would be necessary in any case.

The difficulty with the backwards-difference method arises from the necessity of inverting a matrix at each time step. The actual matrix that must be inverted is the coefficient matrix for the left side of Equations (2.11); the delayed neutrons can be determined directly. Thus, although much larger time steps can be taken with the implicit method than with the explicit method of the previous section, the computation time needed for the matrix inversions may more than offset this advantage. The size time step used in the backwards-difference method is usually limited by the effect of truncation error (of order Δt^2) upon the accuracy of the solution, rather than by numerical stability.

2.3 Implicit Integration—"θ" Method

For a constant $\mathbf{H}$ in the interval $t_p \leqq t \leqq t_{p+1}$, Equation (2.1) has the formal solution

$$\boldsymbol{\Psi}(p+1) = \exp\{\Delta t\, \mathbf{H}\}\boldsymbol{\Psi}(p) = \left\{\mathbf{I} + \Delta t\, \mathbf{H} + \frac{\Delta t^2}{2!}\mathbf{H}^2 + \cdots\right\}\boldsymbol{\Psi}(p). \tag{2.15}$$

The algorithms of Equations (2.2) and (2.9) may be considered as approximations to Equation (2.15). An improved algorithm results from the prescription[3,4]

$$\boldsymbol{\Psi}(p+1) - \boldsymbol{\Psi}(p) = \Delta t\,[\mathbf{M}\boldsymbol{\Psi}(p+1) + (\mathbf{H} - \mathbf{M})\boldsymbol{\Psi}(p)], \tag{2.16}$$

with matrix elements of **M** and **H** related by

$$m_{ij} = \theta_{ij} h_{ij}, \tag{2.17}$$

if the m_{ij}, thus the θ_{ij}, are chosen so that $\mathbf{\Psi}(p+1)$ calculated from Equation (2.16) agrees with $\mathbf{\Psi}(p+1)$ calculated from Equation (2.15). This requires

$$\mathbf{M} = \frac{1}{\Delta t}\mathbf{I} - \mathbf{H}[\exp(\Delta t\,\mathbf{H}) - \mathbf{I}]^{-1}. \tag{2.18}$$

Assuming that **H** has distinct eigenvalues, it may be diagonalized by the transformation

$$(\mathbf{J}^{+})^{T}\mathbf{H}\mathbf{J} = \mathbf{\Gamma} \tag{2.19}$$

where **J** and $\mathbf{J}^{+}$ are the modal matrices corresponding to **H** and $\mathbf{H}^{T}$ (i.e., the columns of **J** and $\mathbf{J}^{+}$ are the eigenvectors of **H** and $\mathbf{H}^{T}$, respectively), and $\mathbf{\Gamma}$ is a diagonal matrix composed of the eigenvalues of **H**. Thus,

$$(\mathbf{J}^{+})^{T}\mathbf{M}\mathbf{J} = \frac{1}{\Delta t}\mathbf{I} - \mathbf{\Gamma}[\exp(\Delta t\,\mathbf{\Gamma}) - \mathbf{I}]^{-1} = \mathbf{L}, \tag{2.20}$$

with **L** diagonal. From this it follows that

$$\mathbf{M} = \mathbf{J}\mathbf{L}(\mathbf{J}^{+})^{T}, \tag{2.21}$$

and the factors θ_{ij} can be determined from

$$\theta_{ij} = \frac{m_{ij}}{h_{ij}}$$

after the m_{ij} are found from Equation (2.21).

Because solving for the θ_{ij} rigorously would entail a great deal of effort, several approximations are made in employing this method to arrive at an algorithm for the solution of the multigroup kinetics equations.[4] The delayed neutrons are treated as sources, and thus are neglected in the determination of the θ_{ij}. An average space-independent value of θ_{ij} is calculated based on a flux square weighting procedure. The delayed neutron precursors have a separate θ_{ij}. Denoting the θ_{ij} associated with groups g and g' as $\theta_{gg'}$, and θ_{ij} associated with the delayed neutrons as θ_d, the following algorithm results:

$$C_m(p+1) = \frac{1-(1-\theta_d)\lambda_m\,\Delta t}{1+\theta_d\lambda_m\,\Delta t}\,C_m(p) + \frac{\Delta t\,\beta_m}{1+\theta_d\lambda_m\,\Delta t}\left[\sum_{g=1}^{G} \nu^g\Sigma_f{}^g(p+1)\,\phi^g(p+1)\,\theta_{1g} + \sum_{g=1}^{G} \nu^g\Sigma_f{}^g(p)\phi^g(p)(1-\theta_{1g})\right], \tag{2.22}$$

$$m = 1, \ldots, M,$$

$$\begin{aligned}
&\theta_{gg}[\nabla\cdot D^g(p+1)\,\nabla\phi^g(p+1) - (\Sigma_a{}^g(p+1)+\Sigma_s{}^g(p+1))\phi^g(p+1)] \\
&+ \sum_{g'\neq g}^{G} \theta_{gg'}\Sigma_s^{g'/g}(p+1)\phi^{g'}(p+1) \\
&+ \chi_P{}^g(1-\beta)\sum_{g'=1}^{G} \theta_{gg'}\nu^{g'}\Sigma_f^{g'}(p+1)\phi^{g'}(p+1) - \frac{1}{\Delta t\,v^g}\,\phi^g(p+1) \\
&+ \sum_{m=1}^{M} \frac{\chi_m{}^g\lambda_m\,\Delta t\,\beta_m\theta_d}{1+\theta_d\lambda_m\,\Delta t}\sum_{g'=1}^{G} \nu^{g'}\Sigma_f^{g'}(p+1)\phi^{g'}(p+1)\theta_{1g'} \\
&= -(1-\theta_{gg})[\nabla\cdot D^g(p)\,\nabla\phi^g(p) - (\Sigma_a{}^g(p)+\Sigma_s{}^g(p))\phi^g(p)] \\
&- \sum_{g'\neq g}^{G} (1-\theta_{gg'})\Sigma_s^{g'/g}(p)\phi^{g'}(p) \\
&- (1-\beta)\chi_P{}^g\sum_{g'=1}^{G} (1-\theta_{gg'})\nu^{g'}\Sigma_f^{g'}(p)\phi^{g'}(p) \\
&- \frac{1}{\Delta t\,v^g}\phi^g(p) - \sum_{m=1}^{M}\frac{\chi_m{}^g\lambda_m C_m(p)}{1+\theta_d\lambda_m\Delta t} - \sum_{m=1}^{M}\frac{\chi_m{}^g\theta_d\lambda_m\,\Delta t\,\beta_m}{1+\theta_d\lambda_m\,\Delta t} \\
&\times \sum_{g'=1}^{G} \nu^{g'}\Sigma_f^{g'}(p)\phi^{g'}(p)(1-\theta_{1g'}),
\end{aligned} \tag{2.23}$$

$$g = 1, \ldots, G.$$

In the limit $\theta_{gg'}, \theta_d \to 1$, Equations (2.22) and (2.23) reduce to the backwards-difference algorithms of Equations (2.10) and (2.11), while Equations (2.22) and (2.23) reduce to the forward-difference algorithms of Equations (2.3) and (2.4) in the limit $\theta_{gg'}, \theta_d \to 0$.

As mentioned, a number of approximations are made in arriving at Equations (2.22) and (2.23), so the mathematical properties associated with Equations (2.16)–(2.21) are not rigorously retained by Equations (2.22) and (2.23). Numerical experience, for one-dimensional, two-group problems, indicates the algorithm of Equations (2.22) and (2.23) is (a) numerically stable for time steps two orders of magnitude greater than are required for stability of Equations (2.3) and (2.4), and (b) is somewhat more accurate than the algorithm of Equations (2.10) and (2.11) for the same time steps. The algorithm of Equations (2.23) requires inversion of the same type of matrix as does the backwards-difference algorithm of Equations (2.11), and, in addition, requires computation of $\theta_{gg'}$ and θ_d, although this latter computation is negligible with respect to the time required for the matrix inversion. In practice, the θ's are predetermined based on experience or intuition.

2.4 Implicit Integration—Time-Integrated Method

Equations (1.2) may, in principle, be integrated directly between t_p and t_{p+1},

$$\begin{aligned} C_m(p+1) &= \exp(-\lambda_m \, \Delta t)\, C_m(p) \\ &\quad + \beta_m \int_{t_p}^{t_{p+1}} dt \exp[-\lambda_m(t_{p+1} - t)] \\ &\quad \times \sum_{g=1}^{G} \nu^g \Sigma_f{}^g(t) \phi^g(t). \end{aligned} \tag{2.24}$$

If the assumption is made that the group-fission rate at each point varies linearly in time in the interval $t_p \leqq t \leqq t_{p+1}$, Equation (2.24) yields an implicit integration algorithm for the precursors,

$$\begin{aligned} C_m(p+1) &= \exp(-\lambda_m \, \Delta t) C_m(p) \\ &\quad + \frac{\beta_m}{\lambda_m} \Bigg[\left\{ \frac{1 - \exp(-\lambda_m \, \Delta t)}{\lambda_m \, \Delta t} - \exp(-\lambda_m \, \Delta t) \right\} \\ &\quad \times \sum_{g=1}^{G} \nu^g \Sigma_f{}^g(p) \phi^g(p) \\ &\quad - \left\{ \frac{1 - \exp(-\lambda_m \, \Delta t)}{\lambda_m \, \Delta t} - 1 \right\} \sum_{g=1}^{G} \nu^g \Sigma_f{}^g(p+1) \phi^g(p+1) \Bigg]. \end{aligned} \tag{2.25}$$

Integration of Equation (1.1) over the interval $t_p \leqq t \leqq t_{p+1}$, with the assumption that all reaction rates vary linearly in that interval, results in an implicit integration algorithm for the neutron flux,

$$\begin{aligned}
&\nabla \cdot D^g(p+1)\,\nabla\phi^g(p+1) - (\Sigma_a{}^g(p+1) + \Sigma_s{}^g(p+1))\phi^g(p+1) \\
&\quad + \sum_{g' \neq g}^{G} \Sigma_s^{g'/g}(p+1)\phi^{g'}(p+1) \\
&\quad + \left[\chi_P{}^g - \sum_{m=1}^{M} \beta_m(\chi_P{}^g - \chi_m{}^g) + \sum_{m=1}^{M} \frac{2}{\Delta t}\frac{\chi_m{}^g\beta_m}{\lambda_m}\right. \\
&\quad \times \left.\left\{\frac{1-\exp(-\lambda_m\,\Delta t)}{\lambda_m\,\Delta t} - 1\right\}\right]\sum_{g'=1}^{G} \nu^{g'}\Sigma_f^{g'}(p+1)\phi^{g'}(p+1) \\
&\quad - \frac{2}{v^g\,\Delta t}\,\phi^g(p+1) \\
&= -\frac{2}{\Delta t}\sum_{m=1}^{M} \chi_m{}^g(1-\exp(-\lambda_m\,\Delta t))C_m(p) - \frac{2}{v^g\,\Delta t}\,\phi^g(p) \\
&\quad - \left[\chi_P{}^g - \sum_{m=1}^{M} \beta_m(\chi_P{}^g - \chi_m{}^g)\right. \\
&\quad - \left.\sum_{m=1}^{M} \frac{2}{\Delta t}\frac{\chi_m{}^g\beta_m}{\lambda_m}\left\{\frac{1-\exp(-\lambda_m\,\Delta t)}{\lambda_m\,\Delta t} - \exp(-\lambda_m\,\Delta t)\right\}\right] \\
&\quad \times \sum_{g'=1}^{G} \nu^{g'}\Sigma_f^{g'}(p)\phi^{g'}(p) - \nabla \cdot D^g(p)\,\nabla\phi^g(p) + (\Sigma_a{}^g(p) + \Sigma_s{}^g(p))\phi^g(p) \\
&\quad - \sum_{g' \neq g}^{G} \Sigma_s^{g'/g}(p)\phi^{g'}(p).
\end{aligned} \tag{2.26}$$

In arriving at Equation (2.26), integration of the precursors was treated as in Equation (2.25); i.e., the group-fission rate was assumed to vary linearly.

Equations (2.25) and (2.26) define the time-integrated algorithm,[5] which, like Equations (2.22) and (2.23), represents an attempt to reduce the truncation error associated with the simple implicit integration formulas of Equations (2.10) and (2.11) without materially increasing the computational time required to obtain a solution. All three implicit integration algorithms require inversion (at each time step) of roughly the same matrix. Limited numerical experience indicates that the θ-method and the time-integrated method yield essentially identical results, and that both methods are somewhat more accurate than the backwards-difference method.

2.5 Perturbation Expansion Method

The time-dependence of the solutions to the kinetics equations is determined by the eigenvalues of the coefficient matrix of the kinetics equations [e.g., by the eigenvalues ω_n of Equation (1.12)]. Algebraically, the largest of these eigenvalues are greater than $-\lambda_1$, where λ_1 is the smallest delayed neutron precursor group decay constant. Algebraically, the smallest of these eigenvalues may be on the order of $-v^g \Sigma_a{}^g$, which can vary from $\sim -10^{+4}$ sec^{-1} for thermal neutrons to $\sim -10^{+8}$ sec^{-1} for fast neutrons. These algebraically smallest eigenvalues are associated with the $\dot{\phi}^g/v^g$ term in the neutron flux equations.

In order for numerical algorithms to describe the time-dependence properly it is necessary to treat these widely separated roots. In the case of the explicit integration method, this required very small time steps to ensure numerical stability of the solution. Numerical stability is not, in general, a problem with the implicit integration methods; however, time steps on the order of $10^2/(v^g \Sigma_a{}^g)$ ($\sim 10^{-2}$ sec for thermal neutrons) may be required for accuracy.

For subcritical and delayed critical transients, the time scale of interest is generally on the order of 10^{-1}–10^1 sec. Thus, the presence of the $\dot{\phi}^g/v^g$ terms in the kinetics equations requires, in the implicit and explicit integration algorithms, the use of time steps that may be orders of magnitude smaller than the time scale of interest.

The difficulty in solving equations containing terms of different orders of magnitude can be eliminated by making a perturbation expansion of the solution in powers of the inverse group velocity and equating terms of like order. This results in a set of equations, each of which contains terms of the same order of magnitude. To illustrate the procedure, Equations (1.3) and (1.4) are written as

$$\mathbf{M}_p \boldsymbol{\phi} + \sum_{m=1}^{M} \chi_m \lambda_m C_m = \varepsilon \mathbf{l} \dot{\boldsymbol{\phi}}, \tag{2.27}$$

$$\beta_m \mathbf{F}^T \boldsymbol{\phi} - \lambda_m C_m = \dot{C}_m, \tag{2.28}$$

$$m = 1, \ldots, M.$$

The quantity ε is the inverse group speed of the lowest energy neutron group. In practice $v^G \ll v^g$, $g < G$. Thus, a valid approximation is to set $v^g = 0$, $g = 1, \cdots, G-1$, in which case $\boldsymbol{l}$ has all zero elements except for a diagonal unit element in the row corresponding to group G. The operator $\mathbf{M}_p$ represents the prompt neutron production and destruction, and may be identified by comparing Equations (2.27) and (1.3).

The flux and precursor densities are expanded,

$$\boldsymbol{\phi} = \boldsymbol{\phi}_0 + \varepsilon\boldsymbol{\phi}_1 + \varepsilon^2\boldsymbol{\phi}_2 + \cdots, \tag{2.29}$$

$$C_m = C_{m0} + \varepsilon C_{m1} + \varepsilon^2 C_{m2} + \cdots. \tag{2.30}$$

When these expansions are substituted into Equations (2.27) and (2.28), and terms of like order in ε are equated, the following equations obtain:

$$\varepsilon^0: \quad \mathbf{M}_p\boldsymbol{\phi}_0 = -\sum_{m=1}^{M} \boldsymbol{\chi}_m \lambda_m C_{m0} \tag{2.31}$$

$$\beta_m \mathbf{F}^T\boldsymbol{\phi}_0 - \lambda_m C_{m0} = \dot{C}_{m0}, \tag{2.32}$$

$$m = 1, \ldots, M,$$

$$\varepsilon^1: \quad \mathbf{M}_p\boldsymbol{\phi}_1 = -\sum_{m-1}^{M} \boldsymbol{\chi}_m \lambda_m C_{m1} + \boldsymbol{l}\dot{\boldsymbol{\phi}}_0, \tag{2.33}$$

$$\beta_m \mathbf{F}^T\boldsymbol{\phi}_1 - \lambda_m C_{m1} = \dot{C}_{m1}, \tag{2.34}$$

$$m = 1, \ldots, M,$$

$$\mathbf{M}_p\dot{\boldsymbol{\phi}}_0 = -\sum_{m=1}^{M} \boldsymbol{\chi}_m \lambda_m \{\beta_m \mathbf{F}^T\boldsymbol{\phi}_0 - \lambda_m C_{m0}\} - \dot{\mathbf{M}}_p\boldsymbol{\phi}_0, \tag{2.35}$$

$$\varepsilon^2: \quad \mathbf{M}_p\boldsymbol{\phi}_2 = -\sum_{m=1}^{M} \boldsymbol{\chi}_m \lambda_m C_{m2} + \boldsymbol{l}\dot{\boldsymbol{\phi}}_1, \tag{2.36}$$

$$\beta_m \mathbf{F}^T\boldsymbol{\phi}_2 - \lambda_m C_{m2} = \dot{C}_{m2}, \tag{2.37}$$

$$m = 1, \ldots, M,$$

$$\mathbf{M}_p\dot{\boldsymbol{\phi}}_1 = -\sum_{m=1}^{M} \boldsymbol{\chi}_m \lambda_m \{\beta_m \mathbf{F}^T\boldsymbol{\phi}_1 - \lambda_m C_{m1}\} - \dot{\mathbf{M}}_p\boldsymbol{\phi}_1 + \boldsymbol{l}\ddot{\boldsymbol{\phi}}_0, \tag{2.38}$$

$$\mathbf{M}_p\ddot{\boldsymbol{\phi}}_0 = -\sum_{m-1}^{M} \boldsymbol{\chi}_m \lambda_m \{\beta_m [\mathbf{F}^T\dot{\boldsymbol{\phi}}_0 + \dot{\mathbf{F}}^T\boldsymbol{\phi}_0]$$

$$- \lambda_m [\beta_m \mathbf{F}^T\boldsymbol{\phi}_0 - \lambda_m C_{m0}]\} - 2\dot{\mathbf{M}}\dot{\boldsymbol{\phi}}_0, \tag{2.39}$$

and so forth.

In arriving at equations for $\dot{\phi}_0, \dot{\phi}_1, \ddot{\phi}_0$, etc., the equations for $\phi_0, \phi_1, \dot{\phi}_0$, etc., were differentiated, and it was assumed that time derivatives of $\mathbf{M}_p$ of order higher than one vanish.

The ε^0 equations, Equations (2.31) and (2.32), are known as the "prompt jump" or "zero lifetime" approximation. Solution of these equations requires integration of the precursor equations (by one of the previously mentioned algorithms) and inversion of the matrix $\mathbf{M}_p$, which is equivalent to the matrix inversions required in the implicit integration algorithms.

If the explicit integration algorithm is used for the precursors, evaluation of the higher order equations does not require any additional matrix inversion. If an implicit integration algorithm is used for the precursor equations, the matrix which must be inverted to solve Equations (2.31), (2.33), (2.36), etc., is slightly different than the matrix which must be inverted to solve Equations (2.35), (2.38), (2.39), etc., and two matrix inversions are required.

An expansion of this type is known to converge when the net multiplication is less than prompt critical (i.e, the reactor is not critical or supercritical on prompt neutrons alone). The zero lifetime equations have been successfully applied in obtaining solutions for subcritical and slightly supercritical transients. Experience with higher order perturbation expansions has been limited, but encouraging. Special care must be taken with initial conditions for the higher order equations.

2.6 Implicit Integration—GAKIN Method

The mathematical properties of this method derive directly from the properties of the spatial finite difference approximation. This approximation is derived from Equation (1.7), which, for a source-free problem, is

$$\dot{\boldsymbol{\theta}} = \mathbf{K}\boldsymbol{\theta}, \tag{2.40}$$

where

$$\boldsymbol{\theta} = \begin{bmatrix} \boldsymbol{\Psi}^1 \\ \vdots \\ \boldsymbol{\Psi}^G \\ \\ \mathbf{d}_1 \\ \vdots \\ \mathbf{d}_M \end{bmatrix}, \tag{2.41}$$

with $\boldsymbol{\Psi}^g$ and $\mathbf{d}_m$ representing $N \times 1$ column vectors of group fluxes and m-type precursor densities, respectively, at each of N spatial mesh points. The matrix $\mathbf{K}$ can be written in terms of $N \times N$ submatrices $\mathbf{K}_{ij}$, which may be identified by comparing Equations (2.42) and (1.9),

$$\mathbf{K} = \begin{bmatrix} \mathbf{K}_{11} & \mathbf{K}_{12} & \mathbf{K}_{13} \cdots & \mathbf{K}_{1,G+M} \\ \mathbf{K}_{21} & \mathbf{K}_{22} & \mathbf{K}_{23} \cdots & \mathbf{K}_{2,G+M} \\ \vdots & \ddots & & \vdots \\ \vdots & & \ddots & \vdots \\ \mathbf{K}_{G+M,1} & \mathbf{K}_{G+M,2} & \cdots & \mathbf{K}_{G+M,G+M} \end{bmatrix}. \tag{2.42}$$

The $N \times N$ matrices $\mathbf{K}_{ij}$ are split,

$$\mathbf{K}_{ii} = \boldsymbol{\Gamma}_{ii} + v^i \mathscr{D}^i, \qquad 1 \leqq i \leqq G, \tag{2.43}$$

where $\mathscr{D}^i$ represents the coupling among mesh points due to the diffusion term (see Section 1.2).

By splitting $\mathbf{K}$ into a matrix $\mathbf{L}$ which contains all the submatrices below the diagonal block, a matrice $\mathbf{U}$ which contains all the submatrices above the diagonal block, and into the block diagonal matrices $\boldsymbol{\Gamma}$ and $\mathbf{D}$,

$$\mathbf{L} = \begin{bmatrix} 0 & 0 & 0 \cdots & 0 \\ \mathbf{K}_{21} & 0 & 0 \cdots & 0 \\ \vdots & \ddots & & \vdots \\ \vdots & & \ddots & \vdots \\ \mathbf{K}_{G+M,1} & \mathbf{K}_{G+M,2} & \cdots & 0 \end{bmatrix} \tag{2.44}$$

$$\mathbf{U} = \begin{bmatrix} 0 & \mathbf{K}_{12} & \mathbf{K}_{13} & \cdots & \mathbf{K}_{1,G+M} \\ 0 & 0 & \mathbf{K}_{23} & \cdots & \mathbf{K}_{2,G+M} \\ \vdots & & \ddots & & \vdots \\ \vdots & & & \ddots & \vdots \\ 0 & & \cdots & & 0 \end{bmatrix} \tag{2.45}$$

$$\boldsymbol{\Gamma} = \begin{bmatrix} \boldsymbol{\Gamma}_{11} & 0 & 0 & \cdots & 0 \\ 0 & \boldsymbol{\Gamma}_{22} & 0 & \cdots & 0 \\ & & \ddots & & \\ 0 \cdots & & \boldsymbol{\Gamma}_{GG} & \cdots & 0 \\ 0 & \cdots & & \mathbf{K}_{G+1,G+1} & 0 \\ \vdots & & & \ddots & \vdots \\ 0 & \cdots & & & \mathbf{K}_{G+M,G+M} \end{bmatrix} \tag{2.46}$$

$$\mathbf{D} = \begin{bmatrix} v^1\mathscr{D}^1 & 0 & 0 & \cdots & 0 \\ 0 & \cdots & v^2\mathscr{D}^2 & 0 & \cdots & 0 \\ \vdots & & \ddots & \ddots & & \vdots \\ 0 & & \cdots & v^G\mathscr{D}^G & \cdots & 0 \\ 0 & & & \cdots & 0 \cdots & 0 \\ \vdots & & & & \ddots & \vdots \\ 0 & & & & \cdots & 0 \end{bmatrix} \tag{2.47}$$

Equation (2.40) may be written

$$\dot{\boldsymbol{\theta}} - \boldsymbol{\Gamma\theta} = (\mathbf{L} + \mathbf{U})\boldsymbol{\theta} + \mathbf{D}\boldsymbol{\theta}. \tag{2.48}$$

This equation may be formally integrated over the interval $t_p \leqq t \leqq t_{p+1}$,

$$\boldsymbol{\theta}(t_{p+1}) = \exp(\Delta t\,\boldsymbol{\Gamma})\boldsymbol{\theta}(t_p) + \int_0^{\Delta t} dt' \exp[(\Delta t - t')\boldsymbol{\Gamma}](\mathbf{L} + \mathbf{U})\boldsymbol{\theta}(t_p + t')$$
$$+ \int_0^{\Delta t} dt' \exp[(\Delta t - t')\boldsymbol{\Gamma}]\mathbf{D}\,\boldsymbol{\theta}(t_p + t'). \tag{2.49}$$

In the first integral of Equation (2.49) the approximation

$$\boldsymbol{\theta}(t_p + t') = \exp(\boldsymbol{\omega} t')\boldsymbol{\theta}(t_p) \tag{2.50}$$

is made, and the second integral is performed with the assumption

$$\boldsymbol{\theta}(t_p + t') = \exp[-\boldsymbol{\omega}(\Delta t - t')]\boldsymbol{\theta}(t_{p+1}). \tag{2.51}$$

In general, $\boldsymbol{\omega}$ is a diagonal matrix. Using Equations (2.50) and (2.51) in Equation (2.49) results in

$$[\mathbf{I} - (\boldsymbol{\omega} - \boldsymbol{\Gamma})^{-1}(\mathbf{I} - \exp[(\boldsymbol{\Gamma} - \boldsymbol{\omega})\,\Delta t])\mathbf{D}]\boldsymbol{\theta}(t_{p+1})$$
$$= [\exp(\boldsymbol{\Gamma}\,\Delta t) + (\boldsymbol{\omega} - \mathbf{I})^{-1}(\exp(\boldsymbol{\omega}\,\Delta t) - \exp(\boldsymbol{\Gamma}\,\Delta t))(\mathbf{L} + \mathbf{U})]\boldsymbol{\theta}(t_p), \tag{2.52}$$

which may be written

$$\boldsymbol{\theta}(t_{p+1}) = \mathbf{A}\boldsymbol{\theta}(t_p). \tag{2.53}$$

If all the diagonal elements of $\boldsymbol{\omega}$ are equal to ω_1, which is the eigenvalue of

$$\mathbf{K}\boldsymbol{\theta}_n = \omega_n \boldsymbol{\theta}_n \tag{2.54}$$

with largest real part, then from Equation (2.48)

$$(\mathbf{L} + \mathbf{U})\boldsymbol{\theta}_1 = (\omega_1 \mathbf{I} - \boldsymbol{\Gamma} - \mathbf{D})\boldsymbol{\theta}_1. \tag{2.55}$$

From the definition of $\mathbf{A}$ (with $\boldsymbol{\omega} = \omega_1 \mathbf{I}$) it can be shown that

$$\mathbf{A}\boldsymbol{\theta}_1 = \exp(\Delta t\, \omega_1)\boldsymbol{\theta}_1 . \tag{2.56}$$

It was shown in Section 1.2 that ω_1 was real and simple, and that $\boldsymbol{\theta}_1$ was positive. For all real values of ω, hence for $\omega = \omega_1$, $\mathbf{A}$ can be shown[6] to be nonnegative, irreducible, and primitive. From the Perron–Frobenius theorem it follows that $\mathbf{A}$ has a simple, real, largest eigenvalue ρ_1 and a corresponding positive eigenvector. The eigenvalue $\rho_1 = \exp(\Delta t\, \omega_1)$ is seen from Equation (2.56) to have a positive eigenvector which is the fundamental mode solution of the kinetics equations, Equation (2.54). If it can be shown that ρ_1 is the largest eigenvalue of $\mathbf{A}$, then Equation (2.56) indicates the asymptotic solution of the integration algorithm of Equation (2.53) is the asymptotic solution of Equation (2.40) for a step change in properties, which shows that the method is unconditionally numerically stable.

The transpose matrix $\mathbf{A}^T$ has the same properties and eigenvalue spectrum as $\mathbf{A}$:

$$\mathbf{A}^T \mathbf{q}_n = \rho_n \mathbf{q}_n . \tag{2.57}$$

By the Perron–Frobenius theorem, $\mathbf{A}^T$ has a real, simple eigenvalue, ρ_k, which is larger than the real part of the other eigenvalues, and the corresponding eigenvector is positive. Premultiplying Equation (2.57) for $n = k$ by $\boldsymbol{\theta}_1{}^T$, premultiplying Equation (2.56) by $\mathbf{q}_1{}^T$, and subtracting yields

$$0 = [\exp\{\Delta t\, \omega_1\} - \rho_1]\boldsymbol{\theta}_1{}^T \mathbf{q}_1 . \tag{2.58}$$

Because $\boldsymbol{\theta}_1$ and $\mathbf{q}_1$ are positive, Equation (2.58) is satisfied only if

$$\exp\{\Delta t\, \omega_1\} = \rho_1$$

is the real eigenvalue. Thus, the method is numerically unconditionally stable.

Inversion of the matrix on the left of Equation (2.52) to obtain $\mathbf{A}$ can be accomplished by the inversion of G $N \times N$ matrices.[6] In practice, an approximation to ω_1 is obtained by an expression of the form

$$\omega_1 = \frac{1}{\Delta t} \ln \frac{\theta_i(t_p)}{\theta_i(t_{p-1})}$$

where i indicates some component or components of the $\boldsymbol{\theta}$ vector, and different values of ω_1 are used in different parts of the reactor[6] (i.e., $\boldsymbol{\omega} \neq \omega_1 \mathbf{I}$).

2.7 Alternating Direction Implicit Method

The implicit integration methods and the perturbation method of the previous sections all reduced to an algorithm for the neutron flux which required the inversion of a matrix at each time step. When the finite-difference spatial approximation is employed, this matrix is $NG \times NG$, where N is the number of mesh points and G is the number of energy groups. In one-dimensional problems, the matrix to be inverted becomes block tridiagonal with $G \times G$ blocks, and inversion can be accomplished by the backwards-elimination/forward-substitution method (Wachspress,[7] p. 26) and requires the inversion of N $G \times G$ matrices. In the *GAKIN* method, this matrix inversion can be accomplished by inverting G $N \times N$ matrices.

However, for multidimensional problems, the matrix inversion associated with the implicit methods poses a formidable and time-consuming task. Alternate formulations of the θ-method and *GAKIN* method have been proposed[9, 10] to reduce the time required for this matrix inversion. Another technique, designed to eliminate this same problem, is the alternating direction implicit (ADI) method. The basis of the ADI method is to make the algorithm implicit for one space dimension at a time and to alternate the space dimension for which the algorithm is implicit. The ideas involved are illustrated by a two-dimensional problem.[2] The equation for the group g neutron flux can be written in the notation of the previous section as

$$\dot{\mathbf{\Psi}}^g = (v^g \mathscr{D}_x{}^g + \tfrac{1}{2}\mathbf{\Gamma}_{gg})\mathbf{\Psi}^g + (v^g \mathscr{D}_y{}^g + \tfrac{1}{2}\mathbf{\Gamma}_{gg})\mathbf{\Psi}^g + \sum_{g' \neq g}^{G} \mathbf{K}_{gg'}\mathbf{\Psi}^{g'} + \sum_{m=1}^{M} \mathbf{K}_{g,G+m}\mathbf{d}_m, \tag{2.59}$$

where the $N \times N$ diffusion matrix $\mathscr{D}^g$ which represents

$$\frac{\partial}{\partial x} D^g \frac{\partial}{\partial x} + \frac{\partial}{\partial y} D^g \frac{\partial}{\partial y}$$

has been separated into $\mathscr{D}_x{}^g$ which represents

$$\frac{\partial}{\partial x} D^g \frac{\partial}{\partial x}$$

and $\mathscr{D}_y{}^g$ which represents

$$\frac{\partial}{\partial y} D^g \frac{\partial}{\partial y}.$$

For the time step t_p to t_{p+1}, an integration algorithm which is implicit in the x-direction and explicit in the y-direction is chosen. First define

$$\begin{aligned} \mathbf{H}_x^g &\equiv v^g \mathscr{D}_x^g + \tfrac{1}{2}\boldsymbol{\Gamma}_{gg}, \\ \mathbf{H}_y^g &\equiv v^g \mathscr{D}_y^g + \tfrac{1}{2}\boldsymbol{\Gamma}_{gg}, \end{aligned} \tag{2.60}$$

then the algorithm is written

$$\begin{aligned} \boldsymbol{\Psi}^g(p+1) - \boldsymbol{\Psi}^g(p) = \Delta t\,[&\mathbf{H}_x^g(p+1)\boldsymbol{\Psi}^g(p+1) + \mathbf{H}_y^g(p)\boldsymbol{\Psi}^g(p) \\ &+ \sum_{g' \neq g}^{G} \mathbf{K}_{gg'}(p)\boldsymbol{\Psi}^{g'}(p) + \sum_{m=1}^{M} \mathbf{K}_{g,G+m}(p)\mathbf{d}_m(p)], \\ &\qquad g = 1, \ldots, G, \end{aligned}$$

or

$$\begin{aligned} (\mathbf{I} - \Delta t\,\mathbf{H}_x^g(p+1))\boldsymbol{\Psi}^g(p+1) = (\mathbf{I} + \Delta t\,&\mathbf{H}_y^g(p))\boldsymbol{\Psi}^g(p) \\ + \Delta t\Bigg[&\sum_{g' \neq g}^{G} \mathbf{K}_{gg'}(p)\boldsymbol{\Psi}^{g'}(p) \\ &+ \sum_{m=1}^{M} \mathbf{K}_{g,G+m}(p)\mathbf{d}_m(p)\Bigg], \\ &g = 1, \ldots, G. \end{aligned} \tag{2.61}$$

For the time step t_{p+1} to t_{p+2}, an algorithm which is implicit in the y-direction is chosen:

$$\begin{aligned} \boldsymbol{\Psi}^g(p+2) - \boldsymbol{\Psi}^g(p+1) = \Delta t\Bigg[&\mathbf{H}_x^g(p+1)\boldsymbol{\Psi}^g(p+1) + \mathbf{H}_y^g(p+2)\boldsymbol{\Psi}^g(p+2) \\ &+ \sum_{g' \neq g}^{G} \mathbf{K}_{gg'}(p+2)\boldsymbol{\Psi}^{g'}(p+2) \\ &+ \sum_{m=1}^{M} \mathbf{K}_{g,G+m}(p+2)\mathbf{d}_m(p+2)\Bigg], \\ &\qquad g = 1, \ldots, G. \end{aligned} \tag{2.62}$$

Using, for the sake of definiteness, the implicit integration formulas of Equation (2.10) for the precursors,

$$\mathbf{d}_m(p+2) = \frac{1}{1 + \lambda_m\,\Delta t}\,\mathbf{d}_m(p+1) + \frac{\beta_m}{1 + \lambda_m\,\Delta t}\sum_{g=1}^{G} \mathscr{F}^g(p+2)\boldsymbol{\Psi}^g(p+2), \tag{2.63}$$

where $\mathscr{F}^g$ is an $N \times N$ diagonal matrix representing $\nu^g \Sigma_f{}^g$ associated with each mesh point. Using Equation (2.63), Equation (2.62) becomes

$$[\mathbf{I} - \Delta t\, \mathbf{H}_y{}^g(p+2)]\mathbf{\Psi}^g(p+2) - \Delta t\Bigg[\sum_{g' \neq g}^{G} \mathbf{K}_{gg'}(p+2)\mathbf{\Psi}^{g'}(p+2)$$
$$+ \sum_{m=1}^{M} \mathbf{K}_{g,G+m}(p+2)\frac{\beta_m}{1+\lambda_m \Delta t}\sum_{g'=1}^{G} \mathscr{F}^{g'}(p+2)\mathbf{\Psi}^{g'}(p+2)\Bigg]$$
$$= (\mathbf{I} + \Delta t\, \mathbf{H}_x{}^g(p+1))\mathbf{\Psi}^g(p+1)$$
$$+ \Delta t \sum_{m=1}^{M} \frac{1}{1+\lambda_m \Delta t}\mathbf{K}_{g,G+m}(p+2)\mathbf{d}_m(p+1), \qquad (2.64)$$
$$g = 1, \ldots, G.$$

The solution proceeds by alternating between the algorithms of Equations (2.61) and (2.64). If there are $N^{1/2}$ mesh points in both the x- and y-directions, the matrices which must be inverted in order to solve Equations (2.61) and (2.64) can be partitioned so that, rather than inverting an $NG \times NG$ matrix, $N^{1/2}$ $N^{1/2}G \times N^{1/2}G$ matrices are inverted. This happens because the matrix to be inverted in Equations (2.61) only couples mesh points in the x-direction, and the matrix to be inverted in Equation (2.64) only couples mesh points in the y-direction. In the case of Equation (2.61), each of the $N^{1/2}G \times N^{1/2}G$ matrices can be further partitioned into G $N^{1/2} \times N^{1/2}$ matrices, because the neutron source terms due to fission, scattering, and precursor decay are treated explicitly in this step. More general algorithms treat these source terms implicitly in both steps.

Very little experience with the ADI method has been obtained for kinetics problems. However, the method has been studied and used for static problems (Wachspress,[7] Chapter 6).

2.8 Discussion

Several numerical integration algorithms for spatially dependent neutron kinetics problems were presented in this chapter. Many variations and combinations of these basic algorithms are possible. Experience with the various methods is insufficient to justify a quantitative comparison among the algorithms. In assessing the appropriate algorithm to use in a kinetics code, many factors not touched upon in this chapter, such as the storage and information retrieval characteristics of the computer to be used, compatibility with the thermal-hydraulic and other feedback equations, and type of transients for which the code will primarily be used, must be considered.

One-dimensional finite-difference solutions and multidimensional synthesis and nodal solutions can be obtained with relatively small amounts of computing time. Much work remains to be done in obtaining efficient multidimensional finite-difference solutions, although initial efforts in this area are encouraging.[9–12] Other numerical integration methods that have been applied successfully to the space-independent neutron kinetics equations may be extended to the space-dependent problem in the future. These methods are reviewed by Hetrick and Weaver.[13]

REFERENCES

1. L. R. Foulke, "A Modal Expansion Technique for Space-Time Reactor Kinetics," Chapter 3, Ph.D. thesis, Department of Nuclear Engineering, Massachusetts Institute of Technology (1966).
2. K. F. Hansen, "A Comparative Review of Two-Dimensional Kinetics Methods," GA-8169, General Atomic (1967).
3. A. F. Henry and A. V. Vota, "WIGL2—A program for the Solution of the One-Dimensional, Two-Group, Space-Time Diffusion Equations Accounting for Temperature, Xenon, and Control Feedback," WAPD-TM-532, Bettis Atomic Power Laboratory (1965).
4. T. A. Porsching and A. F. Henry, "Some Numerical Methods for Problems in Reactor Kinetics," WAPD-T-2130, Bettis Atomic Power Laboratory (1968).
5. W. M. Stacey, Jr. and C. H. Adams, "The Time-Integrated Method: A Quasi-Static Model for Space-Time Neutronics," *Trans. Am. Nucl. Soc.* **10**, 251 (1967).
6. J. B. Andrews and K. F. Hansen, "Numerical Solution of the Time-Dependent Multigroup Diffusion Equations," *Nucl. Sci. Eng.* **31**, 304 (1968).
7. E. L. Wachspress, *Iterative Solutions of Elliptic Systems.* Prentice-Hall, Englewood Cliffs, New Jersey, 1966.
8. M. Clark and K. F. Hansen, *Numerical Methods of Reactor Analysis.* Academic Press, New York, 1964.
9. J. B. Yasinsky and L. A. Hageman, "On the Solution of the Time-Dependent Group Diffusion Equations by an Implicit Time Differenced Iterative Method," *Proc. Conf. Effective Use Computers in Nucl. Ind.* p. 55. CONF-690401, USAEC, Washington, D.C., 1969.

10. W. T. McCormick, Jr., and K. F. Hansen, "Numerical Solution of the Two-Dimensional Time-Dependent Multigroup Equations," *Proc. Conf. Effective Use Computers in Nucl. Ind.* p. 76, CONF-690401, USAEC, Washington, D.C., 1969.
11. R. S. Denning, R. F. Redmond and S. S. Iyer, "A Stable Explicit Finite-Difference Technique for Spatial Kinetics," *Trans. Am. Nucl. Soc.* **12,** 148 (1969).
12. W. H. Reed and K. F. Hansen, "Two-Dimensional Reactor Kinetics," *Trans. Am. Nucl. Soc.* **12,** 234 (1969).
13. D. L. Hetrick and L. E. Weaver, "Some Recent Advances in Reactor Dynamics and Control," in *Proc. Brookhaven Conf. Industrial Needs and Academic Research in Reactor Kinetics*, p. 106. BNL-50117, Brookhaven National Laboratory, 1968.

Chapter 3

VARIATIONAL SYNTHESIS METHODS

The preceding chapters have been devoted to techniques for reducing the coupled set of partial and ordinary differential equations describing the neutron and precursor kinetics to a coupled set of recursive algebraic relations amenable to numerical computation. This general problem can be approached systematically by the application of variational methods. Variational synthesis methods have been developed extensively since their introductions into reactor physics,[1] and have been used to develop several space-time kinetics approximations.[2–6]

By constructing variational principles having stationarity conditions equivalent to the mathematical description of the neutron and precursor kinetics, it is possible to develop approximating equations with a known physical significance. This objective also can be accomplished in developing a modal expansion approximation. In fact, the variational synthesis method is a systematic way to develop a modal expansion approximation.

In this chapter, use of variational synthesis methods is illustrated by a simple example. Then a very general variational principle for the space-dependent multigroup kinetics equations is given. Finally, the derivation of the multichannel space-time synthesis approximation is discussed as an illustration of the more sophisticated approximations which may be derived from the general variational principle.

3.1 A Simple Variational Synthesis Approximation

The basis of the variational method† is to replace the partial differential equations and associated boundary and initial conditions that define the multigroup kinetics approximation with a functional which is stationary about the solution of these equations with the given initial and boundary conditions. The solution is approximated by an expansion in known functions of one or more of the independent variables, and equations for the unknown expansion coefficients are obtained by substituting these expansions into the functional, and then requiring the latter to be stationary with respect to arbitrary variations in the expansion coefficients. This procedure is best illustrated by considering a simple example.

Consider Equation (1.3) with no source and no delayed neutrons,

$$\begin{aligned}[\nabla \cdot \mathbf{D}(r, t) \nabla - \mathbf{R}_a(r, t) - \mathbf{R}_s(r, t) + \mathbf{S}(r, t) \\ + \chi_P \mathbf{F}^T(r, t)]\phi(r, t) = \mathbf{V}^{-1}\dot{\phi}(r, t),\end{aligned} \tag{3.1}$$

with zero-flux boundary conditions

$$\phi(R, t) = 0, \qquad 0 \leqq t \leqq t_f, \tag{3.2}$$

and appropriate initial conditions,

$$\phi(r, 0) = \mathbf{g}(r), \qquad r \in R. \tag{3.3}$$

A variational functional which is stationary about the solution to Equation (3.1), subject to the conditions in Equations (3.2) and (3.3), is

$$\begin{aligned}F(\phi^*, \phi) = \int_R dr \int_0^{t_f} dt\, \phi^{*T}(r, t)\Big[\nabla \cdot \mathbf{D}(r, t) \nabla - \mathbf{R}_a(r, t) - \mathbf{R}_s(r, t) \\ + \mathbf{S}(r, t) + \chi_P \mathbf{F}^T(r, t) - \mathbf{V}^{-1}\frac{\partial}{\partial t}\Big]\phi(r, t).\end{aligned} \tag{3.4}$$

Requiring this functional to be stationary with respect to arbitrary variations in the vector function ϕ^*, ‡

$$\begin{aligned}&\frac{\delta F}{\delta \phi^{*T}}\,\delta\phi^{*T} \\ &= \int_R dr \int_0^{t_f} dt\, \delta\phi^{*T}\left[\nabla \mathbf{D} \nabla - \mathbf{R}_a - \mathbf{R}_s + \mathbf{S} + \chi_P \mathbf{F}^T - \mathbf{V}^{-1}\frac{\partial}{\partial t}\right]\phi = 0.\end{aligned}$$

† In the parlance of the calculus of variations, the "variational method" of reactor physics consists of replacing the Euler equations by their associated functional and then applying a direct or semidirect method. In general, a family of such functionals exists.

‡ By variations with respect to the vector function ϕ^* is meant independent variations with respect to each of the group functions ϕ_g^*.

This can be satisfied identically for arbitrary $\delta\phi^{*T}$ only if

$$\left[\nabla \cdot \mathbf{D}\nabla - \mathbf{R}_a - \mathbf{R}_s + \mathbf{S} + \chi_P \mathbf{F}^T - \mathbf{V}^{-1}\frac{\partial}{\partial t}\right]\phi = 0$$

for all $r \in R$ and $0 \leqq t \leqq t_f$; i.e., only if the function ϕ satisfies Equation (3.1) identically.

Now consider the nature of the function ϕ^*. Require the functional of Equation (3.4) to be stationary with respect to variations in the function ϕ,

$$\frac{\delta F}{\delta\phi}\delta\phi = \int_R dr \int_0^{t_f} dt\, \phi^{*T}\left[\nabla \cdot \mathbf{D}\nabla\delta\phi - \mathbf{R}_a\delta\phi - \mathbf{R}_s\delta\phi + \mathbf{S}\,\delta\phi + \chi_P\mathbf{F}^T\,\delta\phi - \mathbf{V}^{-1}\frac{\partial}{\partial t}\delta\phi\right] = 0. \tag{3.5}$$

The differentiation and variation operators have been assumed to commute. This assumption is valid only if $\delta\phi$ is continuous in space and time; otherwise, the first and last terms in the brackets are undefined.

Limiting the discussion, for the moment, to continuous variations $\delta\phi$, integrating Equation (3.5) by parts yields

$$\begin{aligned}\frac{\delta F}{\delta\phi}\delta\phi = &\int_R dr \int_0^{t_f} dt\, \delta\phi^T\left[\nabla \cdot \mathbf{D}^T\nabla - \mathbf{R}_a{}^T - \mathbf{R}_s{}^T + \mathbf{S}^T + \mathbf{F}\chi_P{}^T + (\mathbf{V}^{-1})^T\frac{\partial}{\partial t}\right]\phi^* \\ &+ \int_0^{t_f} dt\, \phi^{*T}\mathbf{D}\,\nabla\,\delta\phi\cdot\hat{n}\Big|_{r=R} - \int_0^{t_f} dt\,\delta\phi^T\,\mathbf{D}^T\,\nabla\phi^*\cdot\hat{n}\Big|_{r=R} \\ &- \int_R dr\,\phi^{*T}\mathbf{V}^{-1}\,\delta\phi\Big|_{t=0}^{t=t_f} = 0\end{aligned} \tag{3.5a}$$

where $\hat{n}$ denotes the outward unit normal to the surface bounding the reactor.

In order that an equation and a consistent set of auxilliary conditions for the determination of ϕ^* be obtained from Equation (3.5a), the type of allowable variations $\delta\phi$ must be restricted further. Requiring $\delta\phi$ to vanish on the bounding surface $r = R$ for all t, and requiring that $\delta\phi$ vanish for all r at $t = 0$, allows Equation (3.5a) to be satisfied for otherwise arbitrary $\delta\phi$ if ϕ^* satisfies

$$\left[\nabla \cdot \mathbf{D}^T\nabla - \mathbf{R}_a{}^T - \mathbf{R}_s{}^T + \mathbf{S}^T + \mathbf{F}\chi_P{}^T + (\mathbf{V}^{-1})^T\frac{\partial}{\partial t}\right]\phi^* = 0, \tag{3.6}$$

the boundary condition

$$\phi^*(R, t) = 0, \qquad 0 \leqq t \leqq t_f, \tag{3.7}$$

and the final condition

$$\phi^*(r, t_f) = 0, \qquad r \in R. \tag{3.8}$$

This equation and these boundary and final conditions are consistent with the interpretation[7] of ϕ^* as an importance function that describes the importance of a neutron at a given time and position in determining the future neutron population. Equations (3.6)–(3.8) also form a well-posed mathematical problem [i.e., the auxiliary conditions are sufficient to determine a unique solution of Equation (3.6)].

Thus, the variational functional of Equation (3.4) is equivalent to the system of Equations (3.1)–(3.3) and the "dual" system of Equations (3.6)–(3.8), if the allowed variations $\delta\phi$ are required to be continuous in space and time, to vanish on the boundary of the reactor at all times, and to vanish everywhere in the reactor at $t = 0$. It is informative to consider how these restrictions limit the type of variational synthesis approximations that can be derived from the functional of Equation (3.4).

A variational "time synthesis" (see Section 1.3) approximation may be derived from the variational functional of Equation (3.4) by expanding $\boldsymbol{\phi}$ and $\boldsymbol{\phi}^*$ in known functions $\boldsymbol{\Psi}_n$ and $\boldsymbol{\Psi}_n^*$ of position with unknown time-dependent combining coefficients $\mathbf{a}_n$ and $\mathbf{a}_n^*$, respectively:

$$\boldsymbol{\phi}(r, t) = \sum_{n=1}^{N} \boldsymbol{\Psi}_n(r)\mathbf{a}_n(t),$$

$$\boldsymbol{\phi}^*(r, t) = \sum_{n=1}^{N} \boldsymbol{\Psi}_n^*(r)\mathbf{a}_n^*(t). \tag{3.9}$$

Following the notation of Section 1.3, $\boldsymbol{\Psi}_n$ and $\boldsymbol{\Psi}_n^*$ are diagonal matrices of group expansion functions and $\mathbf{a}_n$ and $\mathbf{a}_n^*$ are column matrices of group expansion coefficients.

The expansions of Equation (3.9) are substituted into Equation (3.4) to obtain a reduced functional

$$F(\mathbf{a}_n^*, \mathbf{a}_n) = \sum_{n,n'=1}^{N} \int_R dr \int_0^{t_f} dt\, \mathbf{a}_{n'}^{*T} \boldsymbol{\Psi}_{n'}^{*T} \Big[\nabla \cdot \mathbf{D}\, \nabla - \mathbf{R}_a - R_s + \mathbf{S} + \chi_P \mathbf{F}^T - \mathbf{V}^{-1} \frac{\partial}{\partial t} \Big] \boldsymbol{\Psi}_n \mathbf{a}_n. \tag{3.10}$$

The "time-synthesis" equations are derived from the requirement that the reduced functional of Equation (3.10) is stationary with respect to arbitrary variations in the expansion coefficients $\mathbf{a}_n^*$:

$$\frac{\delta F}{\delta \mathbf{a}_{n'}^{*T}} \delta \mathbf{a}_{n'}^{*T} = \sum_{n=1}^{N} \int_R dr \int_0^{t_f} dt\, \delta \mathbf{a}_{n'}^{*T}\, \mathbf{\Psi}_{n'}^{*T} \Bigg[\nabla \cdot \mathbf{D}\, \nabla - \mathbf{R}_a - \mathbf{R}_s + \mathbf{S}$$

$$+ \chi_P \mathbf{F}^T - \mathbf{V}^{-1} \frac{\partial}{\partial t} \Bigg] \mathbf{\Psi}_n \mathbf{a}_n = 0,$$

$$n' = 1, \ldots, N.$$

In order for this condition to obtain for arbitrary $\delta \mathbf{a}_{n'}^*$, it is necessary that

$$\sum_{n=1}^{N} \Bigg[\Big(\int_R dr\, \mathbf{\Psi}_{n'}^{*T} [\nabla \cdot \mathbf{D}\, \nabla - \mathbf{R}_a - \mathbf{R}_s + \mathbf{S} + \chi_P \mathbf{F}^T] \mathbf{\Psi}_n \Big) \mathbf{a}_n$$

$$- \Big(\int_R dr\, \mathbf{\Psi}_{n'}^{*T} \mathbf{V}^{-1} \mathbf{\Psi}_n \Big) \dot{\mathbf{a}}_n \Bigg] = 0, \tag{3.11}$$

$$n' = 1, \ldots, N.$$

These equations are subject to initial conditions that are obtained from Equation (3.3) by expanding $\boldsymbol{\phi}$ according to Equation (3.9) and requiring Equation (3.3) to be satisfied in a weighted integral sense, with the $\mathbf{\Psi}_n^*$ of Equation (3.9) acting as the weighting functions.

Comparison of Equations (3.11) and (Equations 1.21), when delayed neutrons and an external source are absent in the latter, shows that the two are identical in form. The arbitrary weighting functions $\mathbf{W}_n$ of Equations (1.21) have been replaced by the importance expansion functions $\mathbf{\Psi}_n^{*T}$ in Equations (3.11), but the actual equations and the definition of coefficients are identical. Moreover, since no restrictions have been placed upon the $\mathbf{\Psi}_n^*$ used in Equation (3.9) these functions subsequently may be chosen in the same fashion as the $\mathbf{W}_n$ would be chosen in Equations (1.21).

Equations for the $\mathbf{a}_n^*$ could be obtained, if desired, by requiring the reduced functional to be stationary with respect to arbitrary variations in $\mathbf{a}_n$. The space-time synthesis approximation of Equations (1.29) could be obtained by using, instead of Equations (3.9), expansions in known functions of less than the total number of spatial dimensions, with expansion coefficients that depended upon the remaining spatial dimensions and time. The reduced functional which resulted when these expansions were substituted into Equation (3.4) could be required to be stationary with respect to the $\mathbf{a}_n^{*T}$ to obtain the variational space-time synthesis approximation, which would be identical to Equations (1.29).

The restrictions of $\delta\boldsymbol{\phi}$ discussed previously manifest themselves as restrictions on the expansion functions. To examine this, consider the function $\boldsymbol{\phi}_0$ which satisfies Equations (3.1)–(3.3) and the function $\boldsymbol{\phi}$ constructed according to Equation (3.9) from the solution of Equations (3.11)

$$\boldsymbol{\phi}(r, t) = \boldsymbol{\phi}_0(r, t) + \delta\boldsymbol{\phi}(r, t) = \sum_{n=1}^{N} \boldsymbol{\Psi}_n(r)\mathbf{a}_n(t).$$

Because the variation $\delta\boldsymbol{\phi}$, as well as the exact solution $\boldsymbol{\phi}_0$, are required to be continuous in space and to satisfy the external boundary conditions, each of the expansion functions $\boldsymbol{\Psi}_n$ must be continuous in space and satisfy the external boundary conditions. The requirement that $\boldsymbol{\phi}_0$ and $\delta\boldsymbol{\phi}$ are continuous in time at all spatial locations can be satisfied identically only if the same set of expansion functions are used for all times $0 \leqq t \leqq t_f$. Similar restrictions are encountered for the space-time synthesis approximation.

In some problems it is *a priori* inappropriate to use the same spatial expansion functions in all parts of the reactor and for all times. This motivates the desire to develop synthesis approximations from variational functionals which admit variations $\delta\boldsymbol{\phi}$ that are discontinuous in space and time. This problem is treated in a general fashion in the remainder of this chapter.

3.2 A Variational Functional

For reasons which will be mentioned later, it is convenient to work with the multigroup P_1 equations in place of Equations (1.1). These equations, along with Equations (1.2), can be written in matrix notation as

$$\nabla \cdot \vec{\mathbf{j}}(r, t) + \boldsymbol{\sigma}_R(r, t)\boldsymbol{\phi}(r, t) - (1 - \beta)\boldsymbol{\chi}_P\mathbf{F}^T(r, t)\boldsymbol{\phi}(r, t) - \mathbf{S}_e(r, t)$$
$$- \sum_{m=1}^{6} \boldsymbol{\chi}_m\lambda_m C^m(r, t) = -\boldsymbol{\tau}\dot{\boldsymbol{\phi}}(r, t),$$
$$\nabla\boldsymbol{\phi}(r, t) + 3\boldsymbol{\sigma}_{\mathrm{tr}}(r, t)\vec{\mathbf{j}}(r, t) = 0, \tag{3.12}$$
$$\beta_m\mathbf{F}^T(r, t)\boldsymbol{\phi}(r, t) - \lambda_m C^m(r, t) = \dot{C}^m(r, t), \tag{3.13}$$
$$m = 1, \ldots, 6,$$

ϕ and $\vec{\mathbf{j}}$ are column vectors whose elements are the flux and current, respectively, in each energy group.

χ_P and χ_m are column vectors whose elements are the fission spectrum of prompt and delayed neutrons, respectively, in each energy group.

F is a column vector whose elements are ν times the fission cross section in each energy group.

τ is a diagonal matrix whose elements are the average reciprocal speed in each energy group.

$\boldsymbol{\sigma}_R$ is a square matrix whose diagonal elements are the total cross section minus the isotropic, within-group scattering cross section for each energy group, and whose off-diagonal elements are minus the isotropic scattering cross section between energy groups.

$\boldsymbol{\sigma}_{\text{tr}}$ is a square matrix identical to $\boldsymbol{\sigma}_R$, except the linearly anisotropic scattering cross section is involved rather than the isotropic scattering cross section.

β_m and λ_m are the delay fraction and decay constant, and C^m is the concentration of delayed neutron precursor type m.

Superscript T indicates a transpose matrix (column vector).

$\mathbf{S}_e$ is a column vector whose elements are the external source in each energy group.

* denotes an adjoint of quantity.

Consider a reactor composed of N separate regions R_n, which are bound by surfaces S_n, across which expansion functions may be spatially discontinuous. Let the time interval of interest $t_0 \leqq t \leqq t_f$ be separated into K contiguous intervals $T_k = t_k - t_{k-1}$, such that the expansion functions may be temporally discontinuous at the $K-1$ times t_k.

The problem of obtaining solutions to Equations (3.12) and (3.13), and the equations mathematically adjoint thereto, which (a) satisfy the proper continuity of flux and current interfacial conditions, (b) satisfy the external boundary conditions of vanishing precursors density, $\phi_{S_0} + d\vec{\mathbf{j}}_{S_0} \cdot \hat{n} = 0$, and $\phi_{S_0}^* + d\vec{\mathbf{j}}_{S_0}^* \cdot \hat{n} = 0$, (c) satisfy the initial ($\phi_0 = \mathbf{g}_0$, $C_0{}^m = h_0{}^m$) and final ($\phi_f{}^* = \mathbf{g}_f{}^*$, $C_f^{m*} = h_f{}^*$) conditions, and are continuous in time, may be replaced by the problem of rendering the functional of Equation (3.14) stationary with respect to completely arbitrary variations in the functions ϕ^*, $\vec{\mathbf{j}}^*$, C^{m*}, ϕ, $\vec{\mathbf{j}}$, and C^m. Arbitrary variations within the regions R_n and on the bounding surfaces S_0 and S_n, and within the time intervals T_k as well as at the time interfaces t_k, t_0, and t_f, are admitted:

$$
\begin{aligned}
&S(\phi^*, \vec{\mathbf{j}}^*, C^{m*}, \phi, \vec{\mathbf{j}}, C^m) \\
&\quad = \left\{ \sum_{k=1}^{K} \int_{T_k} dt \sum_{n=1}^{N} \int_{R_n} dr \left[\phi^{*T}[\sigma_R - (1-\beta)\chi_P \mathbf{F}^T]\phi \right.\right. \\
&\qquad - \phi^{*T} \sum_{m=1}^{6} \lambda_m \chi_m C^m + \tfrac{1}{2}(\phi^{*T}\nabla \cdot \vec{\mathbf{j}} - \vec{\mathbf{j}}^T \cdot \nabla\phi^*) \\
&\qquad + \tfrac{1}{2}(\phi^{*T}\tau\dot{\phi} - \phi^T\tau^T\dot{\phi}^*) - \vec{\mathbf{j}}^{*T} \cdot 3\sigma_{\mathrm{tr}}\vec{\mathbf{j}} - \tfrac{1}{2}(\vec{\mathbf{j}}^{*T} \cdot \nabla\phi - \phi^T \nabla \cdot \vec{\mathbf{j}}^*) \\
&\qquad - \sum_{m=1}^{6} C^{m*}\beta_m \mathbf{F}^T\phi + \sum_{m=1}^{6} C^{m*}\lambda_m C^m + \tfrac{1}{2}\sum_{m=1}^{6}(C^{m*}\dot{C}^m - C^m\dot{C}^{m*}) \\
&\qquad \left.\left. - \phi^{*T}\mathbf{S}_e{}^T + \mathbf{S}_e^{*T}\phi \right] \right\} \\
&\qquad + \left\{ \tfrac{1}{2} \sum_{k=1}^{K} \int_{T_k} dt \sum_{n=1}^{N} \int_{S_n} ds\, \hat{n} \cdot \left[\frac{(\phi_{+n}^{*T} + \phi_{-n}^{*T})}{2} (\vec{\mathbf{j}}_{+n} - \vec{\mathbf{j}}_{-n}) \right.\right. \\
&\qquad - \frac{(\vec{\mathbf{j}}_{+n}^T + \vec{\mathbf{j}}_{-n}^T)}{2}(\phi_{+n}^* - \phi_{-n}^*) - \frac{(\vec{\mathbf{j}}_{+n}^{*T} + \vec{\mathbf{j}}_{-n}^{*T})}{2}(\phi_{+n} - \phi_{-n}) \\
&\qquad \left.\left. + \frac{(\phi_{+n}^T + \phi_{-n}^T)}{2}(\vec{\mathbf{j}}_{+n}^* - \vec{\mathbf{j}}_{-n}^*) \right] \right\} \\
&\qquad + \left\{ \tfrac{1}{2} \sum_{k=1}^{K-1} \sum_{n=1}^{N} \int_{R_n} dr \left[\frac{(\phi_{+k}^{*T} + \phi_{-k}^{*T})}{2} \tau(\phi_{+k} - \phi_{-k}) \right.\right. \\
&\qquad - \frac{\phi_{+k}^T + \phi_{-k}^T}{2} \tau^T(\phi_{+k}^* - \phi_{-k}^*) \\
&\qquad + \sum_{m=1}^{6} \frac{(C_{+k}^{m*} + C_{-k}^{m*})}{2}(C_{+k}^m - C_{-k}^m) \\
&\qquad \left.\left. - \sum_{m=1}^{6} \frac{(C_{+k}^m + C_{-k}^m)}{2}(C_{+k}^{m*} - C_{-k}^{m*}) \right] \right\} \\
&\qquad + \left\{ \tfrac{1}{2} \sum_{n=1}^{N} \int_{R_n} dr \left[(\phi_f^{*T} - 2\mathbf{g}_f^{*T})\tau\phi_f + \phi_0^{*T}\tau(\phi_0 - 2\mathbf{g}_0) \right.\right.
\end{aligned}
$$

$$+ \sum_{m=1}^{6} (C_f^{m*} - 2h_f^{m*})C_f{}^m + \sum_{m=1}^{6} C_0^{m*}(C_0{}^m - 2h_0{}^m)\Bigg]\Bigg\}$$

$$+ \Bigg\{\frac{1}{2} \sum_{k=1}^{K} \int_{T_k} dt \int_{S_0} \Bigg[\phi_{S_0}^* \vec{\mathbf{j}}_{S_0} \cdot \hat{n} + \phi_{S_0}^T \vec{\mathbf{j}}_{S_0}^{*} \cdot \hat{n}$$

$$+ 2d(\vec{\mathbf{j}}_{S_0}^{*T} \cdot \hat{n})(\vec{\mathbf{j}}_{S_0} \cdot \hat{n}) + \sum_{m=1}^{6} C_{S_0}^{m*} C_{S_0}^{m} \Bigg] ds \Bigg\}. \tag{3.14}$$

The first three braces of Equation (3.14) represent an integration over the volume of the reactor and the time interval $t_0 \leqq t \leqq t_f$, using Selengut's treatment† of discontinuities. The first term is an integral over the regions R_n and the time intervals T_k within which the expansion functions are continuous. The second and third braces are explicit representations of the contributions of the spatial and temporal discontinuities, respectively. The term $\hat{n}$ is a unit vector normal outward from the surface S_n, and the subscripts $+n$ and $-n$ refer to the positive and negative sides of a surface with respect to the direction of this normal.

† Consider an integration over the domain $0 \leqq \xi \leqq 2L$ of the integrand $V(\xi)\,(du(\xi)/d\xi)$, where V and u are continuous functions of ξ everywhere except at $\xi = L$. Write

$$\int_0^{2L} d\xi\, V(\xi) \frac{du(\xi)}{d\xi} = \lim_{\varepsilon \to 0} \Bigg[\int_0^{L-\varepsilon} d\xi\, V(\xi) \frac{du(\xi)}{d\xi} + \int_{L+\varepsilon}^{2L} d\xi\, V(\xi) \frac{du(\xi)}{d\xi}$$

$$+ \int_{L-\varepsilon}^{L+\varepsilon} d\xi\, V(\xi) \frac{du(\xi)}{d\xi} \Bigg].$$

The first two terms are well defined and correspond to integrals over R_n and T_k in Equation (3.14). The last term is defined in terms of limiting values of V and u as $\xi \to L$ from $\xi > L$ and $\xi < L$,

$$\lim_{\varepsilon \to 0} \int_{L-\varepsilon}^{L+\varepsilon} d\xi \left[\frac{V(L_+) + V(L_-)}{2} \right] \left[\frac{u(L_+) - u(L_-)}{2\varepsilon} \right] = \left[\frac{V(L_+) + V(L_-)}{2} \right]$$

$$\times \Big[u(L_+) - u(L_-) \Big],$$

and corresponds to the terms in the second and third brackets in Equation (3.14).

Note that an approximation of this type for a second derivative is of order $(1/\varepsilon^2)$, and the corresponding integral is infinite in the limit $\varepsilon \to 0$. This is the reason the P_1, rather than the diffusion, equations are used.

The subscripts $+k$ and $-k$ in the third brace refer to values immediately after and before time t_k, respectively.

The fourth brace contains time-boundary terms that allow expansion functions with arbitrary values at t_f and t_0 to be used. The final brace contains terms evaluated on the external surface of the reactor (denoted by subscript S_0), that allow expansion functions to be used that do not satisfy the boundary conditions of zero precursor concentration, and $\phi_{S_0} + d\vec{\mathbf{j}}_{S_0} \cdot \hat{n} = 0$, $\phi^*_{S_0} + d\vec{\mathbf{j}}^*_{S_0} \cdot \hat{n} = 0$.

The conditions which must be satisfied in order that the functional of Equation (3.14) be stationary with respect to arbitrary variations of $\phi^*, \vec{\mathbf{j}}^*, C^{m*}, \phi, \vec{\mathbf{j}}$, and C^m are established.

The condition $(\delta S/\delta \phi^{*T})\, \delta \phi^{*T} = 0$ requires

$$\begin{aligned}
&\nabla \cdot \vec{\mathbf{j}} + [\sigma_R - (1-\beta)\chi_P \mathbf{F}^T]\phi - \sum_{m=1}^{6} \lambda_m \chi_m C^m - \mathbf{S}_e = -\tau\dot{\phi}, \\
&\qquad\qquad (r,t) \in (R_n, T_k), \\
&\phi_{+k} - \phi_{-k} = 0, \qquad r \in R_n \quad \text{(temporal continuity)}, \\
&\hat{n} \cdot (\vec{\mathbf{j}}_{+n} - \vec{\mathbf{j}}_{-n}) = 0, \qquad t \in T_k \quad \text{(spatial continuity)}, \\
&\phi_0 - \mathbf{g}_0 = 0, \qquad r \in R_n \quad \text{(initial condition satisfied)}.
\end{aligned} \tag{3.15}$$

The condition $(\delta S/\delta \vec{\mathbf{j}}^{*T})\, \delta \vec{\mathbf{j}}^{*T} = 0$ requires

$$\begin{aligned}
&\nabla \phi + 3\sigma_{\text{tr}} \vec{\mathbf{j}} = 0, \qquad (r,t) \in (R_n, T_k), \\
&\phi_{+n} - \phi_{-n} = 0, \qquad t \in T_k \quad \text{(spatial continuity)}, \\
&d\vec{\mathbf{j}}_{S_0} \cdot \hat{n} + \phi_{S_0} = 0, \qquad t \in T_k \quad \text{(boundary condition satisfied)}.
\end{aligned} \tag{3.16}$$

The condition $(\delta S/\delta C^{m*})\, \delta C^{m*} = 0$ requires

$$\begin{aligned}
&\beta_m \mathbf{F}^T \phi - \lambda_m C^m = \dot{C}^m, \qquad (r,t) \in (R_n, T_k), \\
&C^m_{+k} - C^m_{-k} = 0, \qquad r \in R_n \quad \text{(temporal continuity)}, \\
&C^m_{S_0} = 0, \qquad t \in T_k \quad \text{(boundary condition satisfied)}, \\
&C_0^m - h_0^m = 0, \qquad r \in R_n \quad \text{(initial condition satisfied)}.
\end{aligned} \tag{3.17}$$

The condition $(\delta S/\delta \boldsymbol{\phi}^T)\,\delta \boldsymbol{\phi}^T = 0$ requires

$$\nabla \cdot \vec{\mathbf{j}}^* + [\boldsymbol{\sigma}_R{}^T - (1-\beta)\mathbf{F}\boldsymbol{\chi}_P{}^T]\boldsymbol{\phi}^* - \sum_{m=1}^{6} \beta_m C^{m*}\mathbf{F} + \mathbf{S}_e{}^* = \boldsymbol{\tau}^T\dot{\boldsymbol{\phi}}^*, \qquad (r, t)\in(R_n, T_k),$$

$$\begin{aligned} \boldsymbol{\phi}^*_{+k} - \boldsymbol{\phi}^*_{-k} &= 0, && r\in R_n \quad \text{(temporal continuity)},\\ \hat{n}\cdot(\vec{\mathbf{j}}^*_{+n} - \vec{\mathbf{j}}^*_{-n}) &= 0, && t\in T_k \quad \text{(spatial continuity)}, \qquad (3.18)\\ \boldsymbol{\phi}_f{}^* - \mathbf{g}_f{}^* &= 0, && r\in R_n \quad \text{(final condition satisfied)}. \end{aligned}$$

The condition $(\delta S/\delta \vec{\mathbf{j}}^T)\,\delta \vec{\mathbf{j}}^T = 0$ requires

$$\begin{aligned} \nabla\boldsymbol{\phi}^* + 3\boldsymbol{\sigma}_{\text{tr}}^T\vec{\mathbf{j}}^* &= 0, && (r, t)\in(R_n, T_k)\\ \boldsymbol{\phi}^*_{+n} - \boldsymbol{\phi}^*_{-n} &= 0, && t\in T_k \quad \text{(spatial continuity)}, \qquad (3.19)\\ d\vec{\mathbf{j}}_{S_0}\cdot\hat{n} + \boldsymbol{\phi}^*_{S_0} &= 0, && t\in T_k \quad \text{(boundary condition satisfied)}. \end{aligned}$$

The condition $(\delta S/\delta C^m)\,\delta C^m = 0$ requires

$$\begin{aligned} \lambda_m\boldsymbol{\chi}_m{}^T\boldsymbol{\phi}^* - \lambda_m C^{m*} &= -\dot{C}^{m*}, && (r, t)\in(R_n, T_k),\\ C^{m*}_{+k} - C^{m*}_{-k} &= 0, && r\in R_n \quad \text{(temporal continuity)},\\ C^{m*}_{S_0} &= 0, && r\in T_k \quad \text{(boundary condition satisfied)}, \qquad (3.20)\\ C^{m*}_f - h^{m*}_f &= 0, && r\in R_n \quad \text{(final condition satisfied)}. \end{aligned}$$

The notation $(r, t)\in(R_n, T_k)$ signifies that the result indicated to the left obtains for spatial points within R_n and for times within T_k.

The temporal continuity of $\vec{\mathbf{j}}$ and $\vec{\mathbf{j}}^*$ follows from the second of Equations (3.15) and the first of Equations (3.16), and from the second of Equations (3.18) and the first of Equations (3.19), respectively.

Thus, functions that render the variational functional stationary with respect to completely arbitrary variations also satisfy the appropriate balance equations and continuity, initial, final, and external boundary conditions of the neutron space-time problem. Since the adjoint problem is a final-value problem, the final conditions on the adjoint flux and precursor may be specified in accordance with the interpretation placed on the adjoint,[7] or chosen to make the stationary value of the functional physically significant.

The stationary value of the functional is

$$S = \sum_{k=1}^{K} \int_{T_K} dt \sum_{n=1}^{N} \int_{R_n} dr\, \mathbf{S}_e^{*T}(r, t)\boldsymbol{\phi}(r, t)$$

$$- \sum_{n=1}^{N} \int_{R_n} dr\, \mathbf{g}_f^{*T}\tau\boldsymbol{\phi}_f - \sum_{n=1}^{N} \int_{R_n} dr \sum_{m=1}^{6} h_f^{m*} C_f{}^m.$$

When $\mathbf{g}_f{}^* = h_f^{m*} = 0$, $\mathbf{S}_e{}^*$ may be chosen as the column vector of group cross sections, and the stationary value of the functional is a reaction rate integrated over space and time. Thus, the functional may be used to develop a modal approximation or to obtain a value for some integral quantity (e.g., power, fission-neutron production) which is accurate to second order with respect to errors in the functions $\boldsymbol{\phi}^*$, $\vec{\mathbf{j}}^*$, C^{m*}, $\boldsymbol{\phi}$, $\vec{\mathbf{j}}$, and C^m.

It should be noted, that in using a variational principle as the basis of a modal expansion method, that the allowed variations in the functions are not completely arbitrary since the dependence upon some of the independent variables is specified. As a consequence, rendering the functional stationary ensures only that the appropriate balance equations and continuity and boundary conditions are satisfied in weighted integral senses. Moreover, the choice of a variational functional is not unique.

3.3 Multichannel Space-Time Synthesis Approximation

For practical reactor models, much of the computational time involved in a modal expansion calculation is associated with computation of the expansion functions. In addition, the efficacy of the calculation depends upon the appropriateness of the expansion functions. Thus, the efficient choice and utilization of expansion functions is an important consideration, and motivates the desire to use different expansion functions in different regions of the reactor or to combine a given set of expansion functions differently in different regions, and to use different sets of expansion functions over different time intervals; i.e., to use expansion functions that are discontinuous in space and time.

The variational functional may be used in a direct manner to derive a quite general set of algebraic equations known as the multichannel space-time synthesis equations. Use of the variational principle implies that the solution of the synthesis equations will (a) satisfy the neutron and precursor balance equations in an importance weighted integral sense within the N regions of the reactor, (b) satisfy continuity of flux and current conditions at all internal surfaces, S_n, in an importance weighted integral sense, and (c) satisfy the appropriate external boundary condition in an importance weighted integral sense.

In order to arrive at specific equations, the reactor is divided conceptually into N channels parallel to the Z-axis (i.e., an XY slice at any axial location is composed of N regions). Each channel is further divided into K axial zones, resulting in a total of $N \times K$ regions. With Z_k denoting the interface between axial zones k and $k+1$, the following axial domain functions are defined:

$$\Delta_k = \begin{matrix} 1, & Z_{k-\theta_k} \geqq Z \geqq Z_{k-1-\theta_{k-1}} \\ 0, & Z_{k-\theta_k} < Z < Z_{k-1-\theta_{k-1}}, \end{matrix}$$

$$\Delta_k{}^* = \begin{matrix} 1, & Z_{k+1-\theta^*_{k+1}} \geqq Z \geqq Z_{k-\theta_k{}^*} \\ 0, & Z_{k+1-\theta_k{}^*} < Z < Z_{k-\theta_k{}^*}, \end{matrix}$$

$$\Delta_k^{1*} = \begin{matrix} 1, & Z_{k-\theta_k^{1*}} \geqq Z \geqq Z_{k-1-\theta^{1*}_{k-1}} \\ 0, & Z_{k-\theta_k^{1*}} < Z < Z_{k-1-\theta^{1*}_{k-1}}, \end{matrix}$$

where the subscripts θ run over the range $0 < \theta < 1$. This implies, for example, that $Z_{k-\theta_k}$ may take on any value in the axial interval $Z_k > Z > Z_{k-1}$, depending upon the choice of θ_k.

Planar domain functions, Δ_n, are defined as unity in channel n and zero elsewhere. These domain functions are assumed to include the bounding surfaces of channel n.

Time domain functions are defined as follows:

$$\Delta_i = \begin{matrix} 1, & t_{i-1} \leqq t \leqq t_i \\ 0, & t_{i-1} > t > t_i, \end{matrix}$$

$$\Delta_i{}^* = \begin{matrix} 1, & t_{i-\theta_i{}^*} \leqq t \leqq t_{i+1-\theta^*_{i+1}} \\ 0, & t_{i-\theta_i{}^*} > t > t_{i+1-\theta^*_{i+1}}, \end{matrix}$$

$$\Delta_{di} = \begin{matrix} 1, & t_{i-\theta_{di}} \leqq t \leqq t_{i+1-\theta_{di+1}} \\ 0, & t_{i-\theta_{di}} > t > t_{i+1-\theta_{di+1}}, \end{matrix}$$

$$\Delta^*_{di} = \begin{matrix} 1, & t_{i-\theta^*_{di}} \leqq t \leqq t_{i+1-\theta^*_{di+1}} \\ 0, & t_{i-\theta^*_{di}} > t > t_{i+1-\theta^*_{di+1}}. \end{matrix}$$

Again, the subscripts θ run over the range $0 < \theta < 1$. These domain functions are used to define a region in space and an interval of time within which a given expansion of the flux, current and precursor densities is made.

Having established this notation, the flux, current, and precursor concentration are expanded in the form

$$\phi(x, y, z, t) = \sum_{n=1}^{N} \Delta_n \sum_{k=1}^{K} \Delta_k \sum_{i=1}^{I} \Delta_i \sum_{\mu=1}^{M} \mathbf{\Psi}_{nki}^{\mu}(x, y)\mathbf{a}_{nki}^{\mu}, \tag{3.21a}$$

$$\begin{aligned} \vec{\mathbf{j}}(x, y, z, t) = \sum_{n=1}^{N} \Delta_n \sum_{k=1}^{K} \Delta_k \sum_{i=1}^{I} \Delta_i \sum_{\mu=1}^{M} [&\mathbf{\Psi}_{nki}^{\mu}(x, y)\mathbf{b}_{nki}^{\mu}(\hat{\imath} + \hat{\jmath}) \\ &- \mathbf{D}_{nki}(x, y)\nabla\mathbf{\Psi}_{nki}^{\mu}(x, y)\,\mathbf{d}_{nki}^{\mu} + \mathbf{\Psi}_{nki}^{\mu}(x, y)\mathbf{g}_{nki}^{\mu}\hat{k}], \end{aligned} \tag{3.21b}$$

$$C^m(x, y, z, t) = \frac{\beta_m}{\lambda_m} \sum_{n=1}^{N} \Delta_n \sum_{k=1}^{K} \Delta_k \sum_{i=1}^{I} \Delta_{di} \sum_{\mu=1}^{M} \mathbf{F}_{nki}^{T}(x, y)\mathbf{\Psi}_{nki}^{\mu}(x, y)\boldsymbol{\pi}\,{}^{m}C_{nki}^{\mu} \tag{3.21c}$$

$$m = 1, \ldots, 6.$$

In Equations (3.21), $\mathbf{\Psi}_{nki}^{\mu}$ is a $G \times G$ diagonal matrix whose elements are the $2D(XY)$ group fluxes for expansion mode μ in channel n and axial zone k during time interval i. The quantities $\mathbf{a}_{nki}^{\mu}$, $\mathbf{b}_{nki}^{\mu}$, $\mathbf{d}_{nki}^{\mu}$, and $\mathbf{g}_{nki}^{\mu}$ are $G \times 1$ column matrices whose elements are the combining coefficients for each energy group of the corresponding expansion function, and ${}^{m}C_{nki}^{\mu}$ is the scalar expansion coefficient associated with $\mathbf{\Psi}_{nki}^{\mu}$ in the expansion of precursor density type m. The gradient operator is two-dimensional $(\hat{\imath}\nabla_x + \hat{\jmath}\nabla_y)$, the diffusion coefficient matrix $\mathbf{D}$ is the inverse of $(3\sigma_{tr})$, and $\boldsymbol{\pi}$ is a $G \times 1$ column matrix whose elements are unity. The subscripts nki on $\mathbf{D}$ and $\mathbf{F}$ denote values in channel n and axial zone k during time interval i. The quantities $\hat{\imath}$, $\hat{\jmath}$, and $\hat{k}$ are unit vectors along the x, y, and z axes. The notation ϕ implies a $G \times 1$ column matrix of group fluxes, and similarly $\vec{\mathbf{j}}$ denotes a $G \times 1$ column matrix of group currents. The precursor density, C^m, is a scalar.

A similar expansion is made for the adjoint quantities:

$$\phi^*(x, y, z, t) = \sum_{n=1}^{N} \Delta_n \sum_{k=1}^{K} \Delta_k{}^* \sum_{i=1}^{I} \Delta_i{}^* \sum_{\mu=1}^{M} \mathbf{\Psi}_{nki}^{*\mu}(x, y)\mathbf{a}_{nki}^{*\mu}, \tag{3.22a}$$

$$\begin{aligned} \vec{\mathbf{j}}^*(x, y, z, t) = \sum_{n=1}^{N} \Delta_n \sum_{k=1}^{K} \Delta_k^{1*} \sum_{i=1}^{I} \Delta_i{}^* \sum_{\mu=1}^{M} [&\mathbf{\Psi}_{nki}^{*\mu}(x, y)\mathbf{b}_{nki}^{*\mu}(\hat{\imath} + \hat{\jmath}) \\ &- \mathbf{D}_{nki}^{*}(x, y)\nabla\mathbf{\Psi}_{nki}^{*\mu}(x, y)\mathbf{d}_{nki}^{*\mu} + \mathbf{\Psi}_{nki}^{*\mu}(x, y)\mathbf{g}_{nki}^{*\mu}\hat{k}], \end{aligned} \tag{3.22b}$$

$$C^{m*}(x, y, z, t) = \sum_{n=1}^{N} \Delta_n \sum_{k=1}^{K} \Delta_k \sum_{i=1}^{I} \Delta_{di}^{*} \sum_{\mu=1}^{M} \boldsymbol{\chi}_m{}^T \boldsymbol{\Psi}_{nki}^{*\mu}(x, y) \pi^m C_{nki}^{*\mu}. \tag{3.22c}$$

Quantities in Equations (3.22) are similar to those discussed with respect to Equations (3.21), except for $\boldsymbol{\chi}_m$ which is a $G \times 1$ column matrix of group delayed neutron emission spectra for precursor type m, and $\mathbf{D}^*$ which is the inverse of $(3\boldsymbol{\sigma}_{\mathrm{tr}}^*)$.

Although a different set of expansion functions $\boldsymbol{\Psi}_{nki}^{\mu}$ can, in principle, be used in each of the $N \times K$ different regions during each time interval, such a procedure would probably never be employed. A more probable procedure is to collect the axial zones into a few axial expansion function regions. Different sets of expansion functions would be used in different expansion function regions. Within a given axial expansion function region, the same set of expansion functions would be used, with different expansion coefficients, in each channel and axial zone. The set of expansion functions may change at a few times during the transient, but would not generally change every time interval.

The synthesis equations are obtained by substituting Equations (3.21) and (3.22) into Equation (3.14) and requiring that the variational functional be stationary with respect to arbitrary variations in each of the adjoint expansion coefficients. (The adjoint synthesis equations are obtained by taking variations with respect to the direct expansion coefficients.)

$$\frac{\delta S}{\delta(\mathbf{a}_{nki}^{*\mu})^T} = 0, \tag{3.23a}$$

$$\frac{\delta S}{\delta(\mathbf{b}_{nki}^{*\mu})^T} = 0, \tag{3.23b}$$

$$\frac{\delta S}{\delta(\mathbf{d}_{nki}^{*\mu})^T} = 0, \tag{3.23c}$$

$$\frac{\delta S}{\delta(\mathbf{g}_{nki}^{*\mu})^T} = 0, \tag{3.23d}$$

$$\frac{\delta S}{\delta^m C_{nki}^{*\mu}} = 0, \tag{3.23e}$$

$$\mu = 1, \ldots, M;$$

$$n = 1, \ldots, N;$$

$$k = 1, \ldots, K;$$
$$i = 1, \ldots, I;$$
$$m = 1, \ldots, 6.$$

In Equations (3.23a)–(3.23d) the variation with respect to the column matrix of adjoint expansion coefficients implies the independent variation with respect to each of the G elements (corresponding to the G energy groups), except when the groups are collapsed, in which case a single expansion coefficient pertains to all G components in the expansion function and the variation is with respect to that single coefficient. Thus, Equations (3.23a)–(3.23d) represent $4MGNKI$ algebraic equations in $(4G+6)MNKI$ unknowns. These equations may be combined to eliminate the **b**, **d**, and **g** coefficients, leaving $MGNKI$ equations, which together with the $6MNKI$ Equations (3.23e), must be solved for the $(G+6)MNKI$ expansion coefficients $\mathbf{a}_{nki}^{\mu}$ and ${}^{m}C_{nki}^{\mu}$.†

The synthesis equations may be written as a recursive relation for the flux combining coefficients:

$$[(\mathbf{P})_i + (\mathbf{A})_i + (\mathbf{R})_i](\mathbf{a})_i = (\mathbf{S})_i, \qquad i = 1, \ldots, I. \tag{3.24}$$

In Equation (3.24), $(\mathbf{P})_i$ is an $MGNK \times MGNK$ block diagonal matrix, with $MG \times MG$ submatrices $(\mathbf{P})_i^{nk}$ corresponding to channel n and axial zone k. The submatrices have scalar elements that are integrals over channel n and axial zone k involving the flux and adjoint flux expansion functions, the prompt and delayed neutron sources, the group removal cross sections, and the inverse neutron group speed. The $MGNK \times MGNK$ block tridiagonal matrix $(\mathbf{A})_i$ describes the axial coupling among regions $(n, k-1)$, (n, k), and $(n, k+1)$ in each channel. The basic scalar matrix elements are integrals over channel n involving the direct and adjoint expansion functions and the elements of the diffusion coefficient matrix.

The planar coupling is described by the $MGNK \times MGNK$ matrix $(\mathbf{R})_i$. The $MGN \times MGN$ submatrices corresponding to each axial zone, k, couple the region (n, k) to all regions (n', k) such that n' is either contiguous to n or is contiguous to a channel n'' which is contiguous to n. This general feature of coupling to "nearest neighbors" and "next nearest neighbors" results for any arbitrary channel structure, as long as the

† At each time interval $(G+6)$ MNK equations must be solved.

domain function Δ_n embraces channel n and its bounding surfaces. The basic scalar elements of the planar coupling matrix are comprised of integrals over the appropriate channels involving the direct and adjoint expansion functions, gradients of these expansion functions, and elements of the diffusion coefficient matrix. In addition, there are surface integrals evaluated on the interfaces separating contiguous channels in an axial zone. These surface integrals involve the direct and adjoint expansion functions and the gradients of these expansion functions.

The $MGNK \times 1$ column matrix $(\mathbf{a})_i$ contains the flux expansion coefficients of Equation (3.21a). The $MGNK \times 1$ column matrix $(\mathbf{S})_i$ depends upon the value of the flux and precursor expansion coefficients in the previous time interval, as well as certain integrals involving the direct and adjoint expansion functions and the various group constants and inverse neutron group speeds.

Equation (3.23e) may be written

$$({}^m\mathbf{C})_{nki} = ({}^m\mathbf{E})_i^{nk}({}^m\mathbf{C})_{nki-1} + ({}^m\boldsymbol{\omega})_i^{nk}(\mathbf{a})_{nki} + ({}^m\boldsymbol{\omega}')_i^{nk}(\mathbf{a})_{nki-1}, \qquad (3.25)$$

$$n = 1, \ldots, N,$$

$$k = 1, \ldots, K,$$

$$i = 1, \ldots, I,$$

$$m = 1, \ldots, 6.$$

The $({}^m\mathbf{C})_{nki}$ are $M \times 1$ column matrices of the precursor expansion coefficients, while $(\mathbf{a})_{nki}$ is an $MG \times 1$ column matrix of the flux expansion coefficients, in channel n and axial zone k. $({}^m\mathbf{E})_i^{nk}$ is an $M \times M$ matrix, while $({}^m\boldsymbol{\omega})_i^{nk}$ and $({}^m\boldsymbol{\omega}')_i^{nk}$ are $M \times MG$ matrices. These matrices have scalar elements which are integrals over the appropriate channel and axial zone involving the direct and adjoint expansion functions and the group-fission cross sections.

The domain functions may be varied to obtain different approximations for the time dependence, the axial coupling among zones, and the planar coupling among channels. The matrices appearing in Equations (3.24) and (3.25) are explicitly presented by Stacey.[8]

A variation of the synthesis technique, in which a single expansion coefficient is assigned to all G group components in a given expansion function, is sometimes used. For this technique, known as collapsed group synthesis (see Section 1.3), the foregoing derivation is applicable, except that the $\boldsymbol{\Psi}_{nki}^{\mu}$ become $G \times 1$ column matrices, the $\mathbf{a}_{nki}^{\mu}$, $\mathbf{b}_{nki}^{\mu}$, $\mathbf{d}_{nki}^{\mu}$,

$\mathbf{g}_{nki}^{\mu}$ become scalars, and the G used in describing the size of the above matrices is replaced by unity.

The multichannel space-time synthesis equations, Equations (3.24) and (3.25), represent a general generic type of approximation. They are not, of course, the only type of approximation that could be derived from the variational functional of Equation (3.14). However, they are sufficiently general to include many of the approximations discussed in the previous two chapters as special cases.

The axial leakage representation and the treatment of the time dependence in Equations (3.24) and (3.25) are essentially finite-difference approximations. A more explicit derivation of spatial finite-difference equations from a variational principle is given by Wachspress.[9]

In that the flux is expanded in known functions and equations are developed for the unknown combining coefficients, the approximation is a generalization of the space-time synthesis method of Section 1.3. When the z dependence and the time dependence of Equations (1.29) and (1.30) are approximated by a suitable finite-difference representation, the resulting equations are identical to Equations (3.24) and (3.25), for the case of only one channel ($N = 1$). Thus, the multichannel space-time synthesis approximation may be considered an extension of the space-time synthesis approximation of Section 1.3, which makes more efficient use of the expansion functions by combining them differently in the different channels of the reactor and which may use different expansion functions in different axial zones and different time intervals. Intuition and numerical experiments[6,8] indicate, for a given set of expansion functions, that the multichannel space-time synthesis approximation is able to represent a wider range of transient flux distributions than is the single channel space-time synthesis approximation of Section 1.3.

A large number of transients have been calculated for a variety of slab geometry models.[8] The purpose of these calculations was to evaluate the extent to which the multichannel feature improved the efficiency of the synthesis calculation, relative to the conventional single channel synthesis.

Single channel synthesis is known to yield quite good results when the expansion functions are specifically tailored to the transient. For example, a transient produced by inserting control rods in a reactor can be calculated well by using the static rodded and unrodded flux shapes as expansion functions. In general, if the expansion functions are selected so that a linear combination of them is capable of representing the flux distribution at any time during a transient, then this transient should be calculated well by the single channel and multichannel synthesis methods using the same expansion functions. In these transients, for which the single channel method produced rather accurate results, the multichannel synthesis results were generally somewhat, but not significantly, better. Thus, the single channel synthesis method is the equal of the multichannel method for a wide range of transients for which it is possible to select expansion functions that are more or less tailored to the specific transient.

However, it is frequently impossible and/or impractical to select expansion functions that are tailored to a specific transient. In transient calculations which include the effect of thermodynamic feedback, and for which control rod motion is determined by a reactor protective system simulator which responds to the neutronic and thermodynamic state of the reactor, it may be impossible to anticipate the range of flux shapes that will be encountered during the transient.

Moreover, in evaluating reactor performance, it is necessary to consider a large number of transients which may involve a wide range of flux distributions.

Although such considerations may, in principle, be mitigated by using enough different expansion functions, this may become impractical for realistic reactor models. The calculation of a detailed planar flux shape may take several hours, even on the fastest of modern digital computers. To calculate a large number of these flux shapes at each of several depletion stages would clearly be prohibitively expensive.

Thus, the multichannel synthesis technique, which makes more efficient use of a given set of expansion functions, is, in principle, superior to the single channel synthesis method. The following example was chosen to provide a concrete illustration of this superiority. The expansion functions were not tailored to the specific transient, and the point to be made is that the multichannel synthesis technique can produce good results in situations for which the single channel synthesis method does poorly. The calculation was based on one group theory with a single group of delayed neutron precursors and employed two expansion functions which were the same for the single channel and multichannel calculations. Time steps of 0.1 sec were employed.

A reflected, three-region slab reactor model was used. The transient was initiated from a critical configuration by linearly decreasing Σ_a in the right quarter of the slab for 0.9 sec. At 0.9 sec, Σ_a in the left quarter of the slab was increased. At 1.0 sec, Σ_a was decreased in the right and left quarters and increased in the central half of the slab. No further

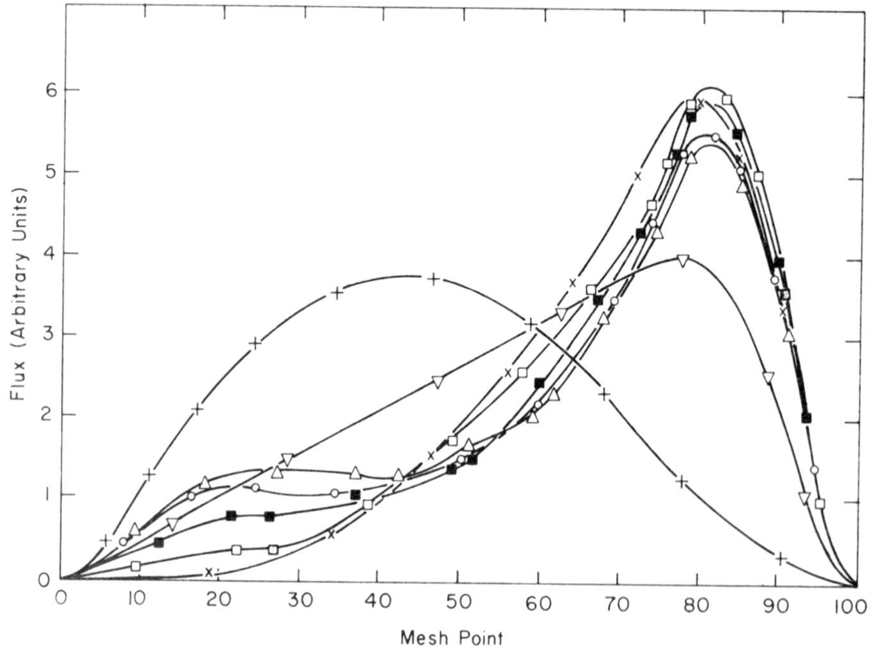

FIGURE 3.1. Flux distribution at 1 sec. KEY: ○, finite-difference (exact); ×, 1 channel with 2 expansion functions; □, 3 channels with 2 expansion functions; ■, 4 channels with 2 expansion functions; △, 5 channels with 2 expansion functions; +, expansion function shape No. 1; ▽, expansion function shape No. 2.

changes in nuclear properties were made. One-, three-, four-, and five-channel synthesis calculations were performed, with the channels taken as (0–100), (0–25,25–75,75–100), (0–25,25–50,50–75,75–100), and (0–25,25–40,40–60,60–75,75–100), respectively. (See Figure 3.1. Reflectors are from 0–5 and 95–100. The three fuel regions are 5–25, 25–75, and 75–95. Each mesh is 1 in.)

The total power calculated with the various synthesis models is compared with the finite-difference RAUMZEIT solution in Figure 3.2. It is apparent that increasing the number of channels significantly improved the calculation of the worth of the perturbations, as manifested in the total power. Increasing the number of channels improved the synthesis calculation of the spatial flux distribution, as shown in Figure 3.1, and, con-

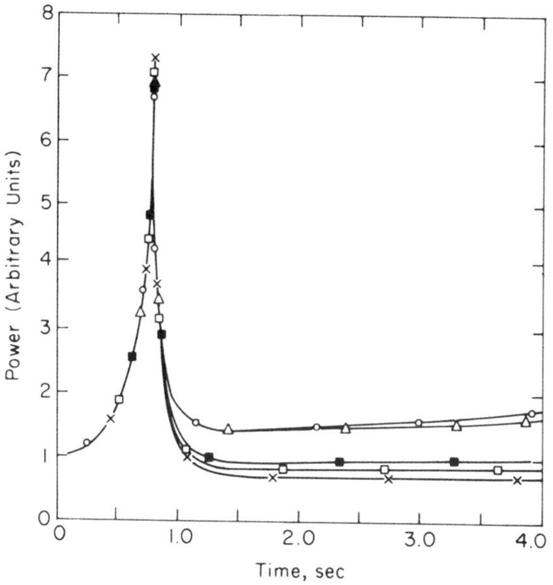

FIGURE 3.2. Total power. KEY: ○, finite-difference; ×, 1 channel; □, 3 channels; ■, 4 channels; △, 5 channels.

sequently, improved the calculation of the worth of the various changes in cross section. In particular, the decrease in Σ_a in the region (5–25) at 1.0 sec had little influence on the single-channel calculation which predicted a negligible neutron flux in this region.

The expansion functions were not tailored to this transient, as is apparent from Figure 3.1, and the single channel synthesis cannot combine them in such a fashion as to obtain an acceptable approximation to the spatial flux distribution during the transient. The multichannel synthesis yields flux distributions from the same expansion functions that become successively better as the number of channels is increased.

Even when only a single channel is used, the formalism of this section has some advantage over the conventional space-time synthesis method presented in Section 1.3 in that axially discontinuous expansion functions may be used; i.e., different sets of expansion functions may be used in different axial regions. As noted in discussing the examples in Chapter 1, axially discontinuous sets of expansion functions may be used in the formalism of Section 1.3 if the interfaces between expansion function regions are treated specially. The variational formalism of this section treats these interfaces automatically and in a manner that is consistent with the treatment of other locations. Alternate treatments of the interface conditions are given by Buslik[10] and Wachspress.[11]

A nodal approximation, as discussed in Section 1.4, is derived by assuming (either explicitly or implicitly) a flux shape, defining weighted average nodal reaction probabilities and internodal coupling based on this flux shape, and developing a set of coupled equations for the flux shape expansion coefficients (or average flux or power) in the different nodes. A similar procedure is followed in deriving Equations (3.24) and (3.25). Consider the region defined by channel n and axial zone k as a node. The multichannel space-time synthesis equations are derived by assuming several flux shapes within each node, defining weighted average reaction probabilities and internodal coupling based on each of these flux shapes, and developing a coupled set of equations relating the expansion coefficients within the different nodes. Thus, the inter- and intranodal neutronics properties of the multichannel approximation can change in time as the combining coefficients of the different expansion functions vary relative to one another. When the time dependence in Equations (1.54) and (1.55) is represented by a suitable finite-difference representation, these equations become identical in form to Equations (3.24) and (3.25), for the case of a single expansion function ($M = 1$), although the coupling terms are defined somewhat differently for the two cases.

REFERENCES

1. D. S. Selengut, "Variational Analysis of a Multidimensional System," p. 89, HW-59126, Hanford Laboratory (1959).
2. D. E. Dougherty and C. N. Shen, "The Space-Time Neutron Kinetics Equations Obtained by the Semidirect Variational Method," *Nucl. Sci. Eng.* **13,** 141 (1962).
3. S. Kaplan, O. J. Marlowe, and J. Bewick, "Application of Synthesis Techniques to Problems Involving Time-Dependence," *Nucl. Sci. Eng.* **18,** 163 (1964).
4. J. B. Yasinsky, "The Solution of the Space-Time Neutron Group Diffusion Equations by a Time Discontinuous Synthesis Method," *Nucl. Sci. Eng.* **29,** 381 (1967).
5. W. M. Stacey, Jr., "Variational Functionals for Space-Time Neutronics," *Nucl. Sci. Eng.* **30,** 448 (1967).

6. W. M. Stacey, Jr., "A Variational Multichannel Space-Time Synthesis Method for Non-Separable Reactor Transients," *Nucl. Sci. Eng.* **34**, 45 (1968).
7. J. Lewins, *Importance: The Adjoint Function.* Pergamon Press, Oxford, 1965.
8. W. M. Stacey, Jr., "A Study of the Multichannel Synthesis Method for Space-Time Neutronics," KAPL-M-6742, Knolls Atomic Power Laboratory (1967).
9. E. L. Wachspress, *Iterative Solution of Elliptic Systems*, Chapter 2. Prentice-Hall, Englewood Cliffs, New Jersey, 1966.
10. A. J. Buslik, "A Variational Principle for the Neutron-Diffusion Equations Using Discontinuous Trial Functions," *Trans. Am. Nucl. Soc.* **9**, 199 (1966).
11. E. L. Wachspress, "On the Use of Different Radial Trial Functions in Different Axial Zones of a Neutron Flux Synthesis Computation," *Nucl. Sci. Eng.* **34**, 342 (1968).

Chapter 4

STOCHASTIC KINETICS

The evolution of the state of a nuclear reactor is essentially a stochastical process, and should, in general, be described mathematically by a set of stochastic kinetics equations. For most problems in reactor physics it suffices to describe the mean value of the state variables in a deterministic manner and to ignore the stochastic aspects. However, the stochastic features of the state variables are important in the analysis of reactor startups in the presence of a weak source,[1–8] and underlie some experimental techniques such as the measurement of the dispersion of the number of neutrons born in fission,[9] the Rossi alpha measurement,[10] and the measurement and interpretation of reactor noise.[11–21]

A stochastic kinetic theory for the space- and energy-independent reactor model has been suggested[22] and studied and elaborated upon in some detail.[2–5,23] Formalisms for a space- and energy-dependent stochastic kinetics theory have been proposed by several authors. A backwards stochastic formalism has received considerable attention,[18,24–26] and an equivalent theory has been developed from consideration of branching processes.[27] Some theories have lumped the relevant physics into Green's functions type kernels that must be obtained from other considerations,[14,20] and formalisms based on the Langevin method have been developed.[19,20] Formalisms based on the forward stochastic method[13,16,17,28] and on the quantum Liouville equation[21] have been suggested.

The purposes of this chapter are to develop a computationally tractable formalism for the calculation of stochastic phenomena in a space- and energy dependent, time-varying, zero-power reactor model, and to present the principal results of a study of these phenomena.[28] The formalism is based on an application of the forward-stochastic method to the neutron-diffusion problem. Alternate formalisms are reviewed also.

4.1 Theory for a Forward Stochastic Model

The spatial domain of a reactor may be partitioned into I space cells, and the energy range of interest may be partitioned into G energy cells. Subject to this partitioning, the state of the reactor is defined by the set of numbers

$$N \equiv \{n_{ig}c_{im}\}, \quad i = 1, \ldots, I; \quad g = 1, \ldots, G; \quad m = 1, \ldots, M,$$

where n_{ig} is the number of neutrons in space cell i and energy cell g, and c_{im} is the number of m-type delayed neutron precursors in space cell i.

Define the transition probability $P(N't' \mid Nt)$ that a reactor that was in state N' at time t' will be in state N at time t. The probability generating function for this transition probability is defined by the relation

$$G(N't' \mid Ut) \equiv \sum_{N} P(N't' \mid Nt) \prod_{igm} u_{ig}^{n_{ig}} v_{im}^{c_{im}}. \tag{4.1}$$

The summation over N implies a summation over all values of n_{ig} and c_{im} for all i, g, and m. The quantities u_{ig} and v_{im} play the role of transform variables.

The transition probability will be written

$$P(N't' \mid Nt) = \Big(\prod_{ig} P_{ig}(N't' \mid n_{ig}t)\Big)\Big(\prod_{im} \hat{P}_{im}(N't' \mid C_{im}t)\Big) \tag{4.2}$$

for mnemonic reasons. This formalism does not denote product probabilities, and is used only to facilitate the distinction between states that differ only by the number of neutrons in one space-energy cell or the number of m-type precursors in one space cell.

Some properties of the probability generating function that will be needed in the subsequent analysis are

$$G(N't' \mid Ut)\Big|_{U=1} = \sum_{N} P(N't' \mid Nt) \equiv 1 \tag{4.3}$$

$$\frac{\partial G}{\partial u_{ig}}(N't' \mid Ut)\Big|_{U=1} = \sum_{N} n_{ig} P(N't' \mid Nt) \equiv \bar{n}_{ig}(t) \tag{4.4}$$

$$\frac{\partial G}{\partial v_{im}}(N't' \mid Ut)\bigg|_{U=1} = \sum_N c_{im} P(N't' \mid Nt) \equiv \bar{c}_{im}(t) \tag{4.5}$$

$$W_{ig,i'g'}(t) \equiv \frac{\partial^2 G(N't' \mid Ut)}{\partial u_{ig}\,\partial u_{i'g'}}\bigg|_{U=1} = \begin{cases} \overline{n_{ig}(t)(n_{ig}(t)-1)}, & ig = i'g', \\ \overline{n_{ig}(t)n_{i'g'}(t)}, & ig \neq i'g'. \end{cases} \tag{4.6}$$

$$Y_{im,i'g'}(t) \equiv \frac{\partial^2 G(N't' \mid Ut)}{\partial v_{im}\,\partial u_{i'g'}}\bigg|_{U=1} = \overline{n_{i'g'}(t)c_{im}(t)} \tag{4.7}$$

$$Z_{im,i'm'}(t) \equiv \frac{\partial^2 G(N't' \mid Ut)}{\partial v_{im}\,\partial v_{i'm'}}\bigg|_{U=1} = \begin{cases} \overline{c_{im}(t)(c_{im}(t)-1)}, & im = i'm', \\ \overline{c_{im}(t)c_{i'm'}(t)}, & im \neq i'm'. \end{cases} \tag{4.8}$$

The notation $U = 1$ indicates that the expression is evaluated for all u_{ig} and v_{im} equal to unity. The overbar denotes an expectation value, as defined explicitly in Equations (4.4) and (4.5). In the foregoing equations, and in the subsequent development, the dependence of the expectation values at time t on the state of the reactor at time t' is implicit.

By considering the events that could alter the state of the reactor during the time interval $t \to t + \Delta t$, balance equations for the transition probability and the probability generating function may be derived. In the limit $\Delta t \to 0$, the probability of more than one event occurring during Δt becomes negligible, and the balance equations can be constructed by summing over all single event probabilities.

(a) Source Neutron Emission

$$\frac{\partial P}{\partial t}\bigg|_S = \sum_{ig} S_{ig}[P_{ig}(n_{ig}-1) - P_{ig}(n_{ig})](\Pi' P_{i'g'})(\Pi \hat{P}_{i'm'}),$$

$$\frac{\partial G}{\partial t}\bigg|_S = \sum_{ig} S_{ig}[u_{ig} - 1]G,$$

(b) Capture Event (includes capture by detectors)

$$\frac{\partial P}{\partial t}\bigg|_c = \sum_{ig} \Lambda_{cig}[(n_{ig}+1)P_{ig}(n_{ig}+1) - n_{ig}P_{ig}(n_{ig})](\Pi' P_{i'g'})(\Pi \hat{P}_{i'm'}),$$

$$\frac{\partial G}{\partial t}\bigg|_c = \sum_{ig} \Lambda_{cig}[1 - u_{ig}]\frac{\partial G}{\partial u_{ig}}.$$

(c) Transport Event

$$\left.\frac{\partial P}{\partial t}\right|_T = \sum_{ig}\sum_{i'} l^g_{ii'}[(n_{ig}+1)P_{ig}(n_{ig}+1)P_{i'g}(n_{i'g}-1)$$

$$- n_{ig}P_{ig}(n_{ig})P_{i'g}(n_{i'g})](\Pi' P_{i''g''})(\Pi \hat{P}_{i''m}),$$

$$\left.\frac{\partial G}{\partial t}\right|_T = \sum_{ig}\sum_{i'} l^g_{ii'}[u_{i'g} - u_{ig}]\,\frac{\partial G}{\partial u_{ig}}.$$

(d) Scattering Event

$$\left.\frac{\partial P}{\partial t}\right|_s = \sum_{ig} \Lambda_{sig}[(n_{ig}+1)P_{ig}(n_{ig}+1)\sum_{g'} K_i^{gg'}P_{ig'}(n_{ig'}-1)(\Pi' P_{i''g''})$$

$$- n_{ig}P_{ig}(n_{ig})(\Pi' P_{i'g'})](\Pi \hat{P}_{i'm'}),$$

$$\left.\frac{\partial G}{\partial t}\right|_s = \sum_{ig} \Lambda_{sig}\left[\sum_{g'} K_i^{gg'}u_{ig'} - u_{ig}\right]\frac{\partial G}{\partial u_{ig}}.$$

(e) Delayed Neutron Emission

$$\left.\frac{\partial P}{\partial t}\right|_d = \sum_{im} \lambda_m[(c_{im}+1)\hat{P}_{im}(c_{im}+1)\sum_{g} \chi_m{}^g P_{ig}(n_{ig}-1)(\Pi' P_{i'g'})$$

$$- c_{im}\hat{P}_{im}(c_{im})(\Pi P_{i'g'})](\Pi' \hat{P}_{i'm'}),$$

$$\left.\frac{\partial G}{\partial t}\right|_d = \sum_{im} \lambda_m\left[\sum_{g} \chi_m{}^g u_{ig} - v_{im}\right]\frac{\partial G}{\partial v_{im}}.$$

(f) Fission Event

$$\left.\frac{\partial P}{\partial t}\right|_f = \sum_{ig} \Lambda_{fig}[(n_{ig}+1)P_{ig}(n_{ig}+1)\sum_{\nu_p} p_g(\nu_p)\{(1-\beta'\nu_p)$$

$$\sum_{g'} \chi_P^{g'} P_{ig'}(n_{ig'} - \nu_p)(\Pi' P_{i''g''})(\Pi \hat{P}_{i'm'})$$

$$+ \sum_{m} \beta_m{}'\nu_p \sum_{g'} \chi_P^{g'} P_{ig'}(n_{ig'} - \nu_p)$$

$$\cdot\, \hat{P}_{im}(c_{im} - 1)(\Pi' P_{i''g''})(\Pi' \hat{P}_{i'm'})\}$$

$$- n_{ig}P_{ig}(n_{ig})(\Pi' P_{i'g'})(\Pi \hat{P}_{i'm'})],$$

$$\left.\frac{\partial G}{\partial t}\right|_{f} = \sum_{ig} \Lambda_{fig}\left[\sum_{g'} f_g \chi_P^{g'} - \{\beta' - \sum_m \beta_m' v_{im}\}\right.$$

$$\left.\cdot \sum_{g'} u_{ig'} \frac{\partial f_g}{\partial u_{ig'}} \chi_P^{g'} - u_{ig}\right] \frac{\partial G}{\partial u_{ig}}.$$

The quantity Λ_{-ig} represents a reaction frequency per neutron, in space cell i and energy cell g, and the subscripts c, s, and f refer to capture, scattering, and fission, respectively.† $K^{gg'}$ is the probability that a scattering event which occurred in energy cell g transfers a neutron to energy cell g', while $\chi_P{}^g$ and $\chi_m{}^g$ are the probabilities that a neutron produced by fission and m-type precursor decay, respectively, has energy within energy cell g. The decay constant for precursor type m is λ_m, and β_m' is the average ratio of the number of m-type precursors to the number of prompt neutrons produced in a fission ($\beta' = \Sigma_m \beta_m'$). S_{ig} is the neutron source rate in space cell i and energy cell g. The quantity $l_{ii'}^g$, represents the frequency per neutron at which neutrons in space cell i and energy cell g will diffuse into space i' (without a change in energy). The prime on the product operator, Π, indicates that the product is taken over all i, g, and m except those explicitly shown in the same term. The quantity f_g is the probability generating function for $p_g(v_p)$, which is the probability distribution function for the number of prompt neutrons emitted in a fission that was caused by a neutron in energy cell g.

$$f_g(u_{ig'}) = \sum_{n_{ig'}} u_{ig'}^{n_{ig'}} p_g(v_p). \tag{4.9}$$

A single fissionable species is assumed for simplicity.

Appropriate balance equations for the transition probability, P, and its probability generating function, G, may be constructed from these terms:

$$\frac{\partial P}{\partial t}(N't' \mid Nt) = \left.\frac{\partial P}{\partial t}\right|_S + \left.\frac{\partial P}{\partial t}\right|_c + \left.\frac{\partial P}{\partial t}\right|_T + \left.\frac{\partial P}{\partial t}\right|_s + \left.\frac{\partial P}{\partial t}\right|_d + \left.\frac{\partial P}{\partial t}\right|_f, \tag{4.10}$$

$$\frac{\partial G}{\partial t}(N't' \mid Ut) = \left.\frac{\partial G}{\partial t}\right|_S + \left.\frac{\partial G}{\partial t}\right|_c + \left.\frac{\partial G}{\partial t}\right|_T + \left.\frac{\partial G}{\partial t}\right|_s + \left.\frac{\partial G}{\partial t}\right|_d + \left.\frac{\partial G}{\partial t}\right|_f. \tag{4.11}$$

† E.g., $\Lambda_{fig} = v^g \Sigma_f{}^g$, v^g = neutron speed; $\Sigma_f{}^g$ = fission cross section.

4.2 Means, Variances, and Covariances

By differentiating Equation (4.11) with respect to u_{ig} and v_{im} and evaluating the resulting expressions for $U = 1$, equations for the mean value of the neutron and precursor distribution, respectively, are obtained [see Equations (4.4) and (4.5)]:

$$\frac{\partial \bar{n}_{ig}(t)}{\partial t} = S_{ig}(t) - \{\Lambda_{cig}(t) + \Lambda_{sig}(t) + \Lambda_{fig}(t)\}\bar{n}_{ig}(t)$$

$$+ \sum_{g'=1}^{G} \Lambda_{sig'}(t) K_i^{g'g} \bar{n}_{ig'}(t) + \chi_P{}^g \sum_{g'=1}^{G} \bar{\nu}_p^{g'} \Lambda_{fig'}(t) \bar{n}_{ig'}(t)$$

$$+ \sum_{m=1}^{M} \chi_m{}^g \lambda_m \bar{c}_{im}(t) + \sum_{i'=1}^{I} l_{i'i}^g(t)(\bar{n}_{i'g}(t) - \bar{n}_{ig}(t)), \tag{4.12}$$

$$\frac{\partial \bar{c}_{im}(t)}{\partial t} = -\lambda_m \bar{c}_{im}(t) + \beta_m{}' \sum_{g=1}^{G} \bar{\nu}_p{}^g \Lambda_{fig}(t) \bar{n}_{ig}(t), \tag{4.13}$$

$$g = 1, \ldots, G; \quad i = 1, \ldots, I; \quad m = 1, \ldots, M.$$

Making use of the identities $\bar{\nu}_p{}^g \equiv (1 - \beta)\bar{\nu}^g$, where $\bar{\nu}^g$ is the average number of neutrons (prompt and delayed) per fission induced by a neutron in energy cell g, and $\beta' = \beta/(1 - \beta)$, it is apparent that these are the conventional space- and energy-dependent neutron and precursor kinetics equations in the finite-difference multigroup approximation [see Equations (1.5) and (1.6)].

By taking second partial derivatives of Equation (4.11) with respect to u_{ig} and v_{im}, and evaluating the result for $U = 1$, equations for the quantities defined by Equations (4.6)–(4.8) are derived:

$$\frac{\partial W_{ig,i'g'}}{\partial t} = S_{ig}\bar{n}_{i'g'} + S_{i'g'}\bar{n}_{ig} - (\Lambda_{cig} + \Lambda_{ci'g'})W_{ig,i'g'}$$

$$+ \sum_j l_{ji}^g (W_{jg,i'g'} - W_{ig,i'g'}) + \sum_j l_{ji'}^{g'} (W_{jg',ig} - W_{jg,ig})$$

$$+ \sum_{g''} [\Lambda_{sig''} K_i^{g''g} W_{i'g',ig} + \Lambda_{si'g''} K_{i'}^{g''g'} W_{i'g'',ig}]$$

$$- (\Lambda_{sig} + \Lambda_{si'g'})W_{ig,i'g'} + \sum_m \lambda_m (\chi_m{}^g Y_{im,i'g'} + \chi_m^{g'} Y_{i'm,ig})$$

$$+ \chi_P^{g} \sum_{g''} \bar{\nu}_p^{g''} \Lambda_{fig''} W_{i'g', ig''} + \chi_P^{g'} \sum_{g''} \bar{\nu}_p^{g''} \Lambda_{fi'g''} W_{ig, i'g''}$$

$$- (\Lambda_{fig} + \Lambda_{fi'g'}) W_{ig, i'g'}$$

$$+ \chi_P^{g} \sum_{g''} \Lambda_{fig''} \overline{\nu_p^{g''}(\nu_p^{g''} - 1)} \bar{n}_{ig''} \, \delta_{ig, i'g'}, \tag{4.14}$$

$$i, i' = 1, \ldots, I; \quad g, g' = 1, \ldots, G.$$

$$\frac{\partial Y_{im, i'g'}}{\partial t} = S_{i'g'} \bar{c}_{im} - \Lambda_{ci'g'} Y_{im, i'g'} + \sum_j l_{ji'}^{g'} (Y_{im, jg'} - Y_{im, i'g'})$$

$$+ \sum_{g''} \Lambda_{si'g''} K_{i'}^{g''g'} Y_{im, i'g''} - \Lambda_{si'g'} Y_{im, i'g'} - \lambda_m Y_{im, i'g'}$$

$$+ \sum_{m'} \lambda_{m'} \chi_{m'}^{g'} Z_{im, i'm'} + \beta_m' \sum_{g''} \bar{\nu}_p^{g''} \Lambda_{fi'g''} W_{i'g', ig''}$$

$$+ \chi_P^{g'} \sum_{g''} \bar{\nu}_p^{g''} \Lambda_{fi'g''} Y_{im, i'g''} - \Lambda_{fi'g'} Y_{im, i'g'}$$

$$+ \beta_m' \sum_{g''} \overline{(\nu_p^{g''})^2} \Lambda_{fi'g''} \bar{n}_{i'g''} \, \delta_{i, i'}, \tag{4.15}$$

$$i, i' = 1, \ldots, I; \quad m = 1, \ldots, M; \quad g' = 1, \ldots, G.$$

$$\frac{\partial Z_{im, i'm'}}{\partial t} = -\lambda_m Z_{im, i'm'} - \lambda_{m'} Z_{im, i'm'} + \beta_m' \sum_g \bar{\nu}_p{}^g \Lambda_{fi'g} Y_{im, i'g}$$

$$+ \beta_{m'}' \sum_g \bar{\nu}_p{}^g \Lambda_{fig} Y_{i'm', ig}, \tag{4.16}$$

$$i, i' = 1, \ldots, I; \quad m, m' = 1, \ldots, M.$$

Equations (4.14)–(4.16) are coupled.

From Equations (4.6)–(4.8) it is apparent that the solutions of Equations (4.14)–(4.16) are related to the variances and covariances of the neutron and precursor distributions; e.g.,

$$\sigma_{ig}^2 \equiv \overline{(n_{ig} - \bar{n}_{ig})^2} = W_{ig, ig} - \bar{n}_{ig}(\bar{n}_{ig} - 1), \tag{4.17}$$

$$\sigma_{im}^2 \equiv \overline{(c_{im} - \bar{c}_{im})^2} = Z_{im, im} - \bar{c}_{im}(\bar{c}_{im} - 1), \tag{4.18}$$

$$\sigma_{igm}^2 \equiv \overline{(n_{ig} - \bar{n}_{ig})(c_{im} - \bar{c}_{im})} = Y_{im, ig} - \bar{n}_{ig} \bar{c}_{im}. \tag{4.19}$$

4.3 Correlation Functions

Define the correlation functions

$$\overline{n_{ig}(t)n_{i'g'}(t')} \equiv \sum_{N}\sum_{N'} n_{ig}n_{i'g'}P(N't' \mid Nt), \tag{4.20}$$

$$\overline{c_{im}(t)n_{i'g'}(t')} \equiv \sum_{N}\sum_{N'} c_{im}n_{i'g'}P(N't' \mid Nt), \tag{4.21}$$

$$\overline{n_{ig}(t)c_{i'm'}(t')} \equiv \sum_{N}\sum_{N'} n_{ig}c_{i'm'}P(N't' \mid Nt), \tag{4.22}$$

$$\overline{c_{im}(t)c_{i'm'}(t')} \equiv \sum_{N}\sum_{N'} c_{im}c_{i'm'}P(N't' \mid Nt). \tag{4.23}$$

By differentiating Equations (4.20)–(4.23) with respect to t, and using Equations (4.3)–(4.6), (4.12), and (4.13), equations satisfied by the correlation functions may be obtained:

$$\begin{aligned}\frac{\partial}{\partial t}\overline{n_{ig}(t)n_{i'g'}(t')} = {} & S_{ig}(t)\bar{n}_{i'g'}(t) - \{\Lambda_{cig}(t) + T_{ig}(t) + \Lambda_{sig}(t) \\ & + \Lambda_{fig}(t)\} \cdot \overline{n_{ig}(t)n_{i'g'}(t')} \\ & + \sum_{m=1}^{M} \lambda_m \chi_m{}^g \overline{c_{im}(t)n_{i'g'}(t')} \\ & + \sum_{g''=1}^{G} \{\Lambda_{sig''}(t)K_i^{g''g}(t) \\ & + \chi_P{}^g \bar{\nu}_p^{g''} \Lambda_{fig''}(t)\} \cdot \overline{n_{ig''}(t)n_{i'g'}(t')},\end{aligned} \tag{4.24}$$

$$\begin{aligned}\frac{\partial}{\partial t}\overline{c_{im}(t)n_{i'g'}(t')} = {} & \beta_m{}' \sum_{g=1}^{G} \bar{\nu}_p{}^g \Lambda_{fig}(t)\overline{n_{ig}(t)n_{i'g'}(t')} \\ & - \lambda_m \overline{c_{im}(t)n_{i'g'}(t)},\end{aligned} \tag{4.25}$$

$$\begin{aligned}\frac{\partial}{\partial t}\overline{n_{ig}(t)c_{i'm'}(t')} = {} & S_{ig}(t)\bar{c}_{i'm'}(t') + \sum_{g''=1}^{G} \{\Lambda_{sig''}(t)K_i^{g''g}(t) \\ & + \chi_P{}^g \bar{\nu}_p^{g''} \Lambda_{fig''}(t)\} \cdot \overline{n_{ig''}(t)c_{i'm'}(t')} \\ & - \{\Lambda_{cig}(t) + T_{ig}(t) + \Lambda_{sig}(t) + \Lambda_{fig}(t)\} \\ & \cdot \overline{n_{ig}(t)c_{i'm'}(t')} + \sum_{m=1}^{M} \lambda_m \chi_m{}^g \overline{c_{im}(t)c_{i'm'}(t')},\end{aligned} \tag{4.26}$$

$$\frac{\partial}{\partial t}\overline{c_{im}(t)c_{i'm'}(t')} = \beta_m{}' \sum_{g=1}^{G} \bar{\nu}_p{}^g \Lambda_{fig}(t)\overline{n_{ig}(t)c_{i'm'}(t')}$$

$$- \lambda_m \overline{c_{im}(t)c_{i'm'}(t')}, \tag{4.27}$$

$$i,\, i' = 1, \ldots, I; \quad m,\, m' = 1, \ldots, M.$$

Equations (4.24) and (4.25) are coupled, as are Equations (4.26) and (4.27). The operator T_{ig} is defined by the operation

$$T_{ig}\overline{n_{ig}n_{i'g'}} = \sum_{i''=1}^{I} l_{i''i}^{g}\left(\overline{n_{i''g}n_{i'g'}} - \overline{n_{ig}n_{i'g'}}\right).$$

4.4 Physical Interpretation, Applications, Initial and Boundary Conditions

If all members of a large ensemble of identical reactors are known to be in an identical state, N', at time t', if all reactors are operated identically subsequent to time t', and if the state, N, of each reactor could be determined at a later time t, the number of reactors in the ensemble that would be found to have a given state, N, would approach the distribution $P(N't' \mid Nt)$. Thus, $\bar{n}_{ig}(t)$ is the ensemble average for the number of neutrons in space cell i and energy cell g at time t, and similarly $\bar{c}_{im}(t)$ is the ensemble average for the number of m-type precursors in space cell i at time t.

Alternately, consider a single reactor that is brought to a known state N' at a reference time t', and subsequently operated in a given manner until a reference time t. If this procedure is repeated a large number of times, the distribution of the number of times the reactor is in a given state N at reference time t approaches $P(N't' \mid Nt)$. Consequently, $\bar{n}_{ig}(t)$ and $\bar{c}_{im}(t)$ are mean values of the neutron and precursor populations, and $\sigma_{ig}^2(t)$ and $\sigma_{im}^2(t)$ are the mean square deviations in these populations. These mean square deviations are an indication of the uncertainty associated with the usual assumption that the actual population is equal to the mean value, which latter is predicted by the conventional kinetics equations. Such considerations are important in analyzing weak-source startup problems.

A set of initial conditions for Equations (4.12)–(4.16) may be obtained from the identity

$$P(N^0 t_0 \mid N t_0) = \delta_{NN^0}, \tag{4.28}$$

where the zero superscript indicates the known state at t_0. From Equations (4.1) and (4.3)–(4.8), the following initial conditions may be deduced:

$$G(N^0 t_0 \mid U t_0) = \prod_{igm} u_{ig}^{n_{ig}^0} u_{im}^{c_{im}^0} \tag{4.29}$$

$$\bar{n}_{ig}(t_0) = \frac{\partial G}{\partial u_{ig}}(N^0 t_0 \mid U t_0)\bigg|_{U=1} = n_{ig}^0 \tag{4.30}$$

$$\bar{c}_{im}(t_0) = \frac{\partial G}{\partial v_{im}}(N^0 t_0 \mid U t_0)\bigg|_{U=1} = c_{im}^0 \tag{4.31}$$

$$W_{ig,\, i'g'}(t_0) = \begin{matrix} n_{ig}^0(n_{ig}^0 - 1), & i'g' = ig, \\ n_{ig}^0 n_{i'g'}^0, & i'g' \neq ig, \end{matrix} \tag{4.32}$$

$$Y_{im,\, i'g'}(t_0) = n_{i'g'}^0 c_{im}^0 \tag{4.33}$$

$$Z_{i'm',\, im}(t_0) = \begin{matrix} c_{im}^0(c_{im}^0 - 1), & i'm' = im, \\ c_{im}^0 c_{i'm'}^0, & i'm' \neq im. \end{matrix} \tag{4.34}$$

In practice, it is not possible to ascertain the "known" initial conditions. This difficulty may be circumvented by using homogeneous initial conditions and, in a subcritical system, taking the asymptotic solution of Equations (4.12)–(4.16) as the initial conditions for further calculations involving changes in operating condition. Alternately, the time-independent versions of Equations (4.12)–(4.16) may be solved to provide initial conditions.

External boundary conditions may be treated by assuming that the space cells on the exterior of the reactor are contiguous to a fictitious external space cell in which the mean value, variance, or covariance is zero, for the purpose of evaluating the net leakage operator. This is equivalent to the familiar extrapolated boundary condition of neutron-diffusion theory.

The interpretation of $P(N't' \mid Nt)$ just discussed leads to an interpretation of the correlation functions. For example, $\overline{n_{ig}(t) n_{i'g'}(t')}$ is the expectation (mean) value of the product of the number of neutrons in space cell i' and energy cell g' at t', and the number of neutrons in space cell i and energy cell g at t. When the reactor properties are time independent, the ensemble average may be replaced by an average over-time in a single reactor (the ergodic theory).† In this case, $\overline{n_{ig}(t' + \tau) n_{i'g'}(t')}$ is amenable

† For a subcritical reactor.

to experimental measurement if the energy and space cells are chosen to conform with the detector resolution. The corresponding theoretical quantity is obtained by solving the time-independent versions of Equations (4.24) and (4.25), using the same type of external boundary treatment discussed before, and employing corrections for the detection process and counting circuit statistics.[29]

4.5 Numerical Studies

Equations (4.12)–(4.16) have been solved numerically for the special case of one energy cell, one delayed neutron precursor type, and one spatial dimension, to study the characteristics of the neutron and precursor distributions under a variety of static and transient conditions. The results of these studies may be characterized in terms of the mean value of the neutron ($\bar{n}_i$) and precursor ($\bar{c}_i$) distributions in region i, and in terms of the relative variances in the neutron and precursor distributions in region i, which are defined by the relations

$$\mu_i \equiv \frac{\overline{(n_i - \bar{n}_i)^2}}{\bar{n}_i^{\,2}} = \frac{W_{i,i} - \bar{n}_i(\bar{n}_i - 1)}{\bar{n}_i^{\,2}} \tag{4.35}$$

$$\varepsilon_i \equiv \frac{\overline{(c_i - \bar{c}_i)^2}}{\bar{c}_i^{\,2}} = \frac{Z_{i,i} - \bar{c}_i(\bar{c}_i - 1)}{\bar{c}_i^{\,2}} \tag{4.36}$$

The quantities μ_i and ε_i are measures of the relative dispersion in the neutron and precursor statistical distributions in region i.

Certain general trends emerge from the numerical studies that have been performed.

(1) When the reactor is subcritical, the asymptotic values of μ_i and ε_i vary from region to region, and within a given region $\varepsilon_i \ll \mu_i$.

(2) When the reactor is subcritical, the asymptotic values of μ_i and ε_i depend upon the source level and distribution and the degree of subcriticality. In general, increasing the source level or the multiplication factor reduces μ_i and ε_i.

(3) When the reactor is supercritical, μ_i and ε_i attain asymptotic values that are identical in all regions, and $\mu_i = \varepsilon_i$.

(4) When the reactor is brought from a subcritical to a supercritical configuration, μ_i generally decreases and ε_i generally increases.

(5) The asymptotic value of μ_i and ε_i in a supercritical reactor is sensitive to the manner in which the reactor is brought supercritical.

(a) For the withdrawal of a single rod (or group of rods) between fixed limits, the more rapid the withdrawal the larger the asymptotic value of μ_i and ε_i.

(b) When a number of rods are to be withdrawn, each rod at the same rate, withdrawing the rods on one side of the reactor and then withdrawing the rods on the other side of the reactor results in a larger asymptotic value for μ_i and ε_i than if all the rods are withdrawn simultaneously.

(c) Withdrawing a rod (group of rods) from position a to position c, then reinserting it (them) to position b ($a > b > c$) results in a larger asymptotic value of μ_i and ε_i than if the rod (group of rods) was withdrawn at the same rate from position a to position b.

(6) The time at which μ_i and ε_i obtain an asymptotic value may differ from region to region, particularly if flux tilting is significant.

(7) When the reactor is brought from a subcritical to a supercritical configuration, the asymptotic value of μ_i and ε_i depends upon the source level and the initial subcritical multiplication factor.

(8) The more supercritical the configuration obtained before μ_i and ε_i attain their asymptotic value, the larger this asymptotic value is.

(9) For a supercritical reactor, μ_i and ε_i generally attain their asymptotic value when $\bar{n}_i$ is of the order of 10^5 n/cm^3.

One of the models studied is depicted in Figure 4.1. This initially subcritical model was made supercritical in four different ways. In transient

Region 1	Region 2	Region 3
D = 1.5 cm	D = 0.1 cm	D = 1.5 cm
Σ_f = 0.008 cm^{-1}	Σ_f = 0.003 cm^{-1}	Σ_f = 0.008 cm^{-1}
Σ_c = 0.0125 cm^{-1}	Σ_c = 0.005 cm^{-1}	Σ_c = 0.0125 cm^{-1}
S = 5 × 10^2 sec^{-1}	S = 0	S = 5 × 10^2 sec^{-1}
← 100 cm →	← 100 cm →	← 100 cm →

FIGURE 4.1. Spatially dependent model ($\beta = 0.0075$, $\lambda = 0.075$, $\nu_p = 2.41$, $\overline{\nu_p(\nu_p - 1)} = 3.84$).

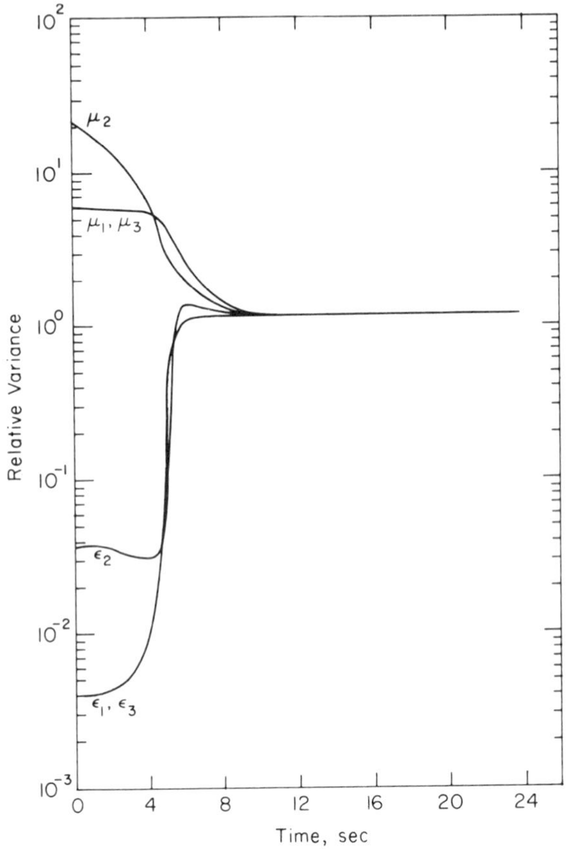

FIGURE 4.2. Relative variances for spatially dependent problem, transient No. 1.

No. 1 Σ_a was linearly decreased in Regions 1 and 3 at a rate of 2.83–4 cm^{-1}/sec for 5 sec. In transient No. 2 Σ_a was linearly decreased in Region 1 at a rate of 2.83–4 cm^{-1}/sec for 5 sec. Transient No. 3 was identical to transient No. 2 for the first 5 seconds, then Σ_a was linearly decreased in Region 3 at a rate of 2.83–4 cm^{-1}/sec from 5–10 sec, resulting in the same final nuclear properties that were attained in transient No. 1. In transient No. 4 Σ_a was linearly decreased in regions 1 and 3 at a rate of 2.83–4 cm^{-1}/sec for 6 sec, then Σ_a was linearly increased in Regions 1 and 3 at a rate of 2.83–4 cm^{-1}/sec for 1 sec, resulting in the same final nuclear

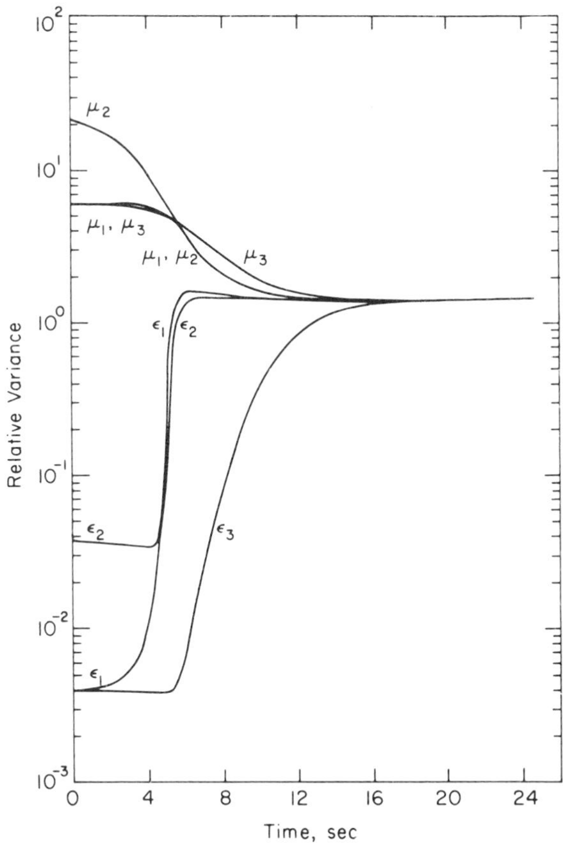

FIGURE 4.3. Relative variances for spatially dependent problem, transient No. 2.

properties that were attained in transients No. 1 and No. 3. The final configuration was slightly less than prompt critical in each transient, and the mean values of the neutron ($\bar{n}_i$) and precursor ($\bar{c}_i$) distributions eventually increased on an asymptotic period. Severe flux tilting (i.e., $\overline{n_1} \gg \overline{n_3}$) occurred during transients No. 2 and No. 3, although the final relation among $\overline{n_1}$, $\overline{n_2}$, and $\overline{n_3}$ was identical for transients No. 1, 3, and 4.

The variation in time of μ_i and ε_i are shown in Figures 4.2–4.5. Transients No. 1, 3, and 4, in which the reactor was brought from a given initial configuration to a given final configuration by three different control sequences, exhibit different loci of μ_i and ε_i as well as different asymptotic values of

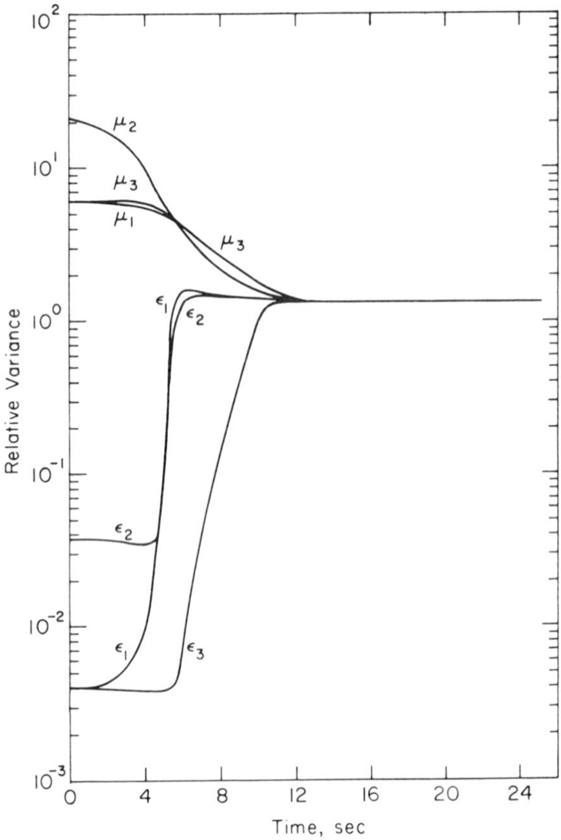

FIGURE 4.4. Relative variances for spatially dependent problem, transient No. 3.

μ_i and ε_i. In transient No. 3, the flux tilting was much greater than in transient No. 1, and in transient No. 4 a configuration more supercritical than the final configuration that was attained.

Comparing Figures 4.3 and 4.4, it is apparent that the effect of decreasing Σ_a in Region 3 during the interval 5–10 sec (in transient No. 3) has a small effect on the loci of μ_i and ε_i, except for ε_3. Thus, changes made in the early stages of the transient, when the mean neutron density ($\bar{n}_i$) was of the order 10–100, were much more influential upon μ_i and ε_i than were later changes which occurred when the mean neutron density was of the order 10^2–10^4. This decrease in Σ_a had a dramatic effect in reducing the

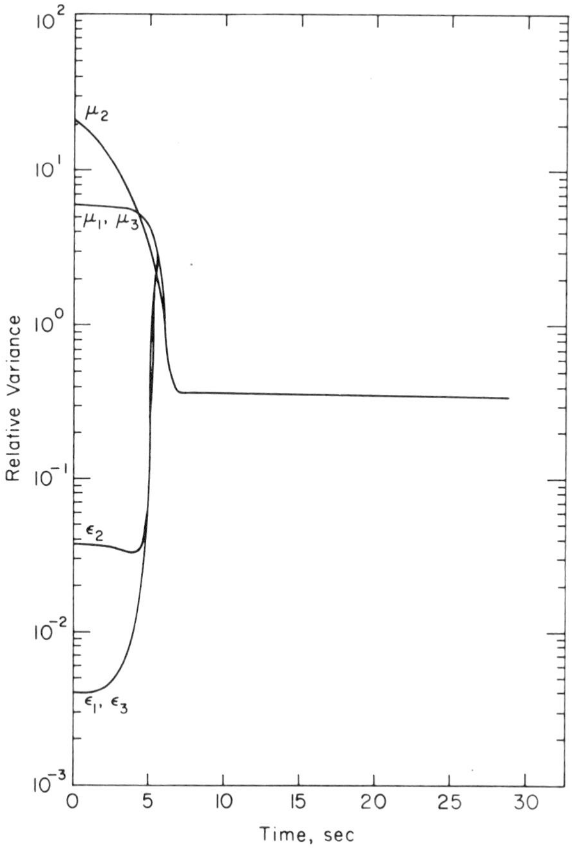

FIGURE 4.5. Relative variances for spatially dependent problem, transient No. 4.

spatial flux tilt, however. This particular reactor model was very loosely coupled, and the observed weak dependence of μ_i and ε_i in one part of the reactor on changes in nuclear properties in other parts of the reactor was increased in more tightly coupled models.

Equations (4.12)–(4.16) have also been solved for the special case of one space-cell and one delayed neutron precursor type. The three-energy cell model with properties indicated in Table 4.1 has been employed to study stochastic phenomena associated with the energy dependent problem. Again, the results may be characterized in terms of the means $\bar{n}_g$ and $\bar{c}$,

TABLE 4.1

MODEL PARAMETERS FOR ENERGY DEPENDENT PROBLEMS

Group (g)	DB^2 (cm^{-1})	$\Sigma_{c+}\Sigma_{f+}\Sigma_s^{g/g+1}$ (cm^{-1})	Σ_f (cm^{-1})	$\Sigma_s^{g/g+1}$ (cm^{-1})	$\overline{\nu_p}$	$\overline{\nu_p(\nu_p-1)}$	v (cm/sec)	χ_P
1	1.435–3	1.918–2	1.19–3	1.56–2	2.60	4.686	4.4+6	1.0
2	0·982–3	1.626–1	6.61–3	1.34–1	2.44	3.958	3.1+5	0.
3	0.846–3	1.167–1	6.01–2	—	2.41	3.840	2.2+5	0.

and relative variances μ_g and ε, of the neutron and precursor distributions, where

$$\mu_g \equiv \frac{\overline{(n_g - \bar{n}_g)^2}}{\bar{n}_g^{\,2}} = \frac{W_{g,g} - \bar{n}_g(\bar{n}_g - 1)}{\bar{n}_g^{\,2}}, \tag{4.37}$$

$$\varepsilon \equiv \frac{\overline{(c - \bar{c})^2}}{\bar{c}^2} = \frac{Z - \bar{c}(\bar{c} - 1)}{\bar{c}^2}, \tag{4.38}$$

The reactor model depicted in Table 4.1 is subcritical. Two calculations were made, both simulating the introduction of a source $S = 10^3$ nt/sec into the fast group (Group 1) of reactors with negligible initial neutron and precursor concentrations. In one calculation, the source was allowed to remain in the reactor indefinitely, whereas in the second calculation the source was removed after 0.125 sec. Delayed neutron parameters $\lambda = 0.075$/sec and $\beta = 0.0075$ were used.

The quantities μ_g of Equation (4.37) are plotted in Figure 4.6 for the case in which the source is removed after 0.125 sec. The initial rapid variation in μ_g occurs over the period in which the source neutrons introduced into Group 1 are slowing down to Groups 2 and 3 to establish an asymptotic spectrum. At $\sim$0.1 sec, the mean neutron densities ($\bar{n}_g$) and the relative variances (μ_g) attain quasiasymptotic values (delayed neutrons are not yet asymptotic). When the source is removed, at 0.125 sec, the neutron densities drop and the relative variances increase, both attaining quasiasymptotic values by 0.25 sec. For both 0.1 and 0.25 secs, $\bar{n}_3 > \bar{n}_2 > \bar{n}_1$ and $\mu_1 > \mu_2 > \mu_3$. When the transient was repeated, but the source was not removed, the quasiasymptotic behavior at 0.1 sec was continued out to 0.25 sec and beyond. In both cases, the asymptotic values of μ_g and $\bar{n}_g$,

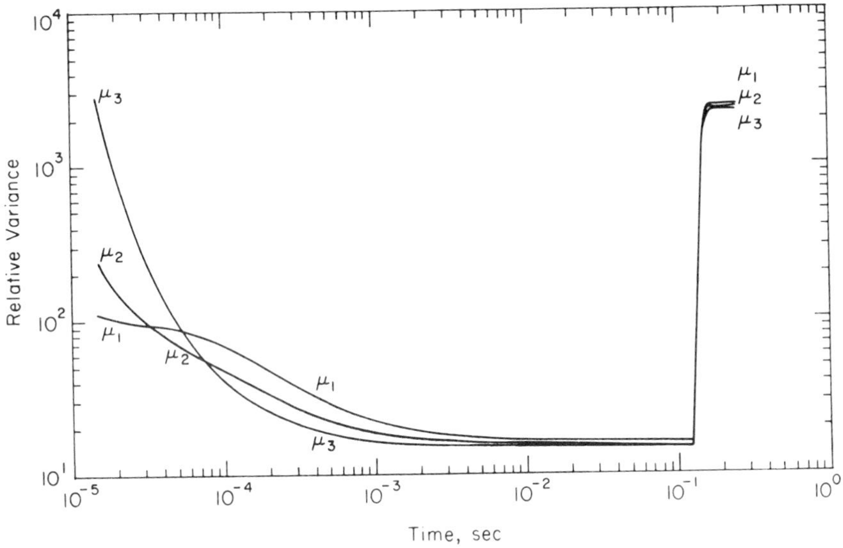

FIGURE 4.6. Relative variances for energy dependent problem.

which attained once the delayed neutron precursors reached an asymptotic distribution, were only slightly different from the values at 0.25 sec.

The same pair of transients was repeated on a reactor model which was made supercritical by increasing Σ_{f3} to 6.41–2 cm^{-1}. The initial variation of the μ_g, associated with the establishment of an asymptotic spectrum, was qualitatively similar to that shown in Figure 4.6. However, the asymptotic values of the μ_g were identical (i.e., $\mu_1 = \mu_2 = \mu_3$ asymptotically), and asymptotically, $\mu_g = \varepsilon$. Removal of the source had no effect on either the mean values or relative variances of the neutron and precursor distribution.

These results appear plausible on a physical basis. In a subcritical reactor, the neutron fluctuations† are governed by the fluctuations of the

† To define what is meant by fluctuations it is necessary to return to the concept of an ensemble of similar reactors. In each reactor the neutron and precursor densities are, in principle, well defined at any instant of time. However, if the neutron or precursor density at a given location and time were observed in a single reactor, this density might differ from the ensemble-average value for the density. Alternately, if a reactor is subcritical and stationary, the neutron density at a certain location may vary in time about a time-average value. Such differences from the mean are the fluctuations referred to in the text.

neutron sources, which are the instantaneous natural and neutron-induced fission rates and delayed neutron precursor decay rate, as well as by the fluctuations of the fission, capture, and diffusion processes. The precursor fluctuations are governed by an integral of the fission fluctuations over several mean lifetimes for the precursors ($\tau_{\text{mean}} = \lambda^{-1}$). This integral dependence of the precursor fluctuations on the fluctuations in the fission process tends to smooth out the fluctuations in the former relative to fluctuations in the latter.

$$C_m(r, t) = \int_{t-n\tau}^{t} dt'\, e^{-\lambda(t-t')} \beta \Sigma_f(r, t') n(r, t') v, \qquad n \sim 10.$$

In a supercritical reactor the precursor fluctuations still depend on an integral of the fission fluctuations over the last few mean precursor lifetimes. However, the major contribution to the integral now comes from times close to the upper limit of the integral. Thus, the precursor fluctuations tend to depend upon the instantaneous fission fluctuations. In a supercritical reactor the major source of prompt neutrons very quickly becomes the neutron-induced fission rate. Thus, the neutron and precursor fluctuations are governed by fluctuations in the instantaneous fission rates, and it is plausible that these fluctuations are statistically identical.†

In a subcritical reactor in which the relative fission and the capture and diffusion probabilities vary from region to region, it is reasonable to expect the fluctuations in the neutron population to exhibit different statistical characteristics from region to region. Similarly, when the relative absorption and scattering probabilities and the fission spectrum differ for the various energy groups in a subcritical reactor, the fluctuations in the neutron populations in the different energy groups plausibly exhibit different statistical characteristics. It is interesting that in a supercritical reactor the fluctuations in the neutron population exhibit asymptotically the same statistical characteristics at all spatial positions and in all energy groups.

From the numerical results, the behavior of the stochastic distribution of the neutron and precursor populations within a reactor can be induced. In subcritical reactors the stochastic neutron distribution is spatially and energy dependent, and the stochastic precursor distribution is spatially dependent. In general, in a subcritical reactor, the stochastic neutron

† Have the same relative variances.

distribution is more disperse than the stochastic precursor distribution at the same spatial location.

In a supercritical reactor, the asymptotic stochastic neutron distribution is space and energy independent and is identical to the asymptotic stochastic precursor distribution. As a reactor is brought from a subcritical to a supercritical configuration, the stochastic neutron distribution generally becomes less disperse, whereas the stochastic precursor distribution becomes more disperse. The dispersion of the asymptotic distribution in a supercritical reactor depends upon the manner in which the reactor attains its final configuration as well as on the multiplicative properties of the initial and final configurations and the source level. The dispersion of the asymptotic distribution is more sensitive to changes that are made to the reactor configuration when the mean neutron and precursor densities are small than to later changes made in the presence of larger mean neutron and precursor densities.

4.6 Relevance to Safeguards Analysis

The essential problem of the safeguards analysis of a reactor startup is the determination of the probability that the actual neutron population is within a prescribed band about the mean neutron population predicted by the deterministic kinetics equations. As a specific example, consider a startup excursion which is terminated by a power level trip actuating the scram mechanism. The scram is initiated at a finite time after the trip point is reached, during which time interval the neutron density continues to increase. If the startup procedure consists of shimming out control rods, the principal concern is that the actual neutron population is less than the mean population, in which case the neutron density at which the trip point is reached occurs later, with the reactor being more supercritical and thus on a shorter period than is predicted by the deterministic kinetics equations. Consequently, the power excursion is more severe than would be predicted deterministically.

Startup analyses may be separated into two phases, stochastic and deterministic. The first phase is analyzed with stochastic kinetics, and the results are used as initial conditions, with associated probabilities, for the second phase, which is analyzed with deterministic kinetics. Feedback effects may be ignored generally during the stochastic phase. A reasonable time to switch from the stochastic to the deterministic phase is the time at which the neutron and precursor distributions obtain their asymptotic shape. This time may probably be approximated by the time at which μ_i and ε_i of Equations (4.35) and (4.36) attain their asymptotic value. If the neutron and precursor distributions [i.e., $P(N't'|Nt_s)$] were known at the switchover time t_s, the probability that the actual neutron and precursor densities are less than some specified values could be calculated.

The asymptotic neutron and precursor distributions in a reactor with large multiplication and no feedback can be approximated by the gamma distribution, which is completely characterized by the mean and variance of the distribution (i.e., by $\bar{n}_i$ and μ_i or $\bar{c}_i$ and ε_i). Use of the gamma distribution is suggested theoretically by the fact that the stationary probability distribution of a variate in a stationary multiplicative process approaches a gamma distribution as the multiplication increases without limit, and is justified empirically by the fact that its use in conjunction with a point reactor kinetics model leads to results that are in reasonable agreement with the GODIVA weak-source transient data.[6]

The gamma distribution is

$$F(x)\,dx = \frac{r^r}{\Gamma(r)} x^{r-1} e^{-rx}\,dx, \tag{4.39}$$

where Γ is the gamma function,[30] x is the ratio of the actual value of the variate to the mean value of the variate, and r is the ratio of the mean value of the variate to the square root of the variance. For example,

$$x = \frac{n_i}{\bar{n}_i}, \qquad r = \mu_i^{-1/2} \tag{4.40}$$

for the monoenergetic model.

From Equation (4.39), the probability that $x < \Delta$ can be computed.

$$\text{Prob}\{x < \Delta\} = \int_0^\Delta F(x)\,dx = 1 - \frac{\Gamma_{\text{in}}(r, \Delta r)}{\Gamma(r)}, \tag{4.41}$$

where Γ_{in} is the incomplete gamma function.[30] This can be written entirely in terms of tabulated functions by using certain identities,

$$\text{Prob}\{x < \Delta\} = \frac{(\Delta r)^r e^{-\Delta r} M(1, r+1, \Delta r)}{r\Gamma(r)}, \tag{4.42}$$

where M is the confluent hypergeometric function.[30]

Based on the results of the stochastic phase, initial conditions for the deterministic phase can be assigned from

$$n_i^0 = \Delta\bar{n}_i(t_s), \qquad c_i^0 = \Delta\bar{c}_i(t_s), \tag{4.43}$$

where $\bar{n}_i$ and $\bar{c}_i$ are the mean values of the neutron and precursor densities at the switchover time, t_s. For a given value of Δ, Equation (4.42) yields the probability that $n_i(t_s) < \Delta\bar{n}_i(t_s)$, $c_i(t_s) < \Delta\bar{c}_i(t_s)$.

4.7 Spatial Stochastic Effects

In order to evaluate the influence of spatial effects upon the stochastic distribution, transients No. 1 and 2 (on the spatially dependent model) were repeated on an equivalent point reactor model. The point reactor model was equivalent to the corresponding spatially dependent models in the sense that it employed the same delayed neutron parameters and had the same initial reactivity. The reactivity insertion for the point model was a 5-sec ramp which led to the same asymptotic period obtained with the corresponding spatially dependent model, so that the effect of spatial flux tilting on the reactivity was accounted for. Thus, the point model was equivalent deterministically to the corresponding spatially dependent model in that both models predicted identical mean values of the total neutron and precursor distribution at all times during the transient.

For transient No. 1, in which negligible flux tilting occurs, both the point and spatially dependent models yielded an asymptotic value of the relative variances of 1.14. For transient No. 2, in which significant flux tilting occurs, the point model predicted an asymptotic value of the relative variance of 0.70, whereas the spatially dependent model predicted an asymptotic value of 1.41. The point model predicts that the asymptotic distribution is less disperse in transient No. 2 than in transient No. 1 (note the reactivity addition is less for the former than for the latter). On the other hand, the spatially dependent model predicts that the asymptotic distribution is more disperse in transient No. 2 than in transient No. 1, with large flux tilting in the former. Thus, apparently there are intrinsic spatial stochastic effects which are not accounted for by a point model, even if that point model reproduces the deterministic features of the transient.

A physical argument can be made in support of the more disperse distribution predicted by the spatially dependent model in transient No. 2.

In this transient, the relative fission and capture probabilities were different in regions 1 and 3. Thus, in addition to the probability that a neutron at a given position could be captured, cause a fission, or leak from the system, the probability that it could diffuse to a region with different relative probabilities must be considered. This additional probablity, which can be incorporated into the spatially dependent model, but not into the point reactor model, increases the uncertainty inherent in the stochastic process and thus increases the dispersion in the stochastic distribution. In transient No. 1, the relative probabilities in regions 1 and 3 were identical, and the likelihood of a neutron being in region 2 was quite small,† so that the probability of a neutron diffusing to a region of different relative probabilities was very small, and the spatially dependent and point models afforded substantially identical representations of the processes involved.

A quantitative measure of the significance of this spatial stochastic effect can be obtained by evaluating Equation (4.42), which yields the probability that the ratio of the actual to mean neutron or precursor population in the reactor is less than some number Δ. This probability, evaluated for transients No. 1 and 2, is shown in Table 4.2. The point and

TABLE 4.2

Prob $\{x < \Delta\}$: PROBABILITY THAT THE RATIO OF THE ACTUAL TO MEAN NUMBER OF NEUTRONS OR PRECURSORS IN THE REACTOR IS LESS THAN Δ

	Prob $\{x < \Delta\}$			
	Transient number 1		Transient number 2	
Δ	Pt. mod.	Spat. mod.	Pt. mod.	Spat. mod.
10^{0}	6.36–1	6.36–1	6.21–1	6.43–1
10^{-1}	1.07–1	1.07–1	6.74–2	1.27–1
10^{-2}	1.29–2	1.29–2	4.55–3	1.89–2
10^{-3}	1.49–3	1.49–3	2.92–4	2.73–3
10^{-6}	2.32–6	2.32–6	7.59–8	8.12–6
10^{-8}	3.10–8	3.10–8	3.09–10	1.68–7

† Region 2 is primarily a decoupling region, and $\bar{n}_2 \ll \bar{n}_1$ or $\bar{n}_3$.

spatially dependent models yield identical results for transient No. 1. However, for transient No. 2, the point model predicts significantly smaller probabilities than are predicted by the spatially dependent model.

Assume that in a safeguards analysis of a startup the initial conditions for the deterministic phase are to be chosen according to Equation (4.43), with Δ selected such that $\text{Prob}\{x < \Delta\} = 10^{-7}$. For transient No. 2, the point model predicts $\Delta \cong 10^{-5}$, but the spatially dependent model predicts $\Delta \cong 10^{-8}$. Thus, the initial conditions for the deterministic phase of the startup analysis are overpredicted by a factor of 10^3 by the point model.

Alternately, assume that the initial conditions of the deterministic phase of a startup analysis are to be evaluated from Equation (4.43) with $\Delta \cong 10^{-6}$. The probability that the actual neutron and precursor concentrations at the beginning of the deterministic phase were less than the initial conditions chosen from Equation (4.43) is $\sim 10^{-7}$ according to the point model, but is $\sim 10^{-5}$ according to the spatially dependent model.

This example indicates that the use of a point model in the safeguards analysis of a reactor with nonuniform properties can be significantly in error, and in a nonconservative manner.

4.8 Theory for a Backwards Stochastic Model

The theory for a backwards stochastic model could be developed analogous to the theory of Section 4.1 by considering the events that could alter the state of the reactor during the time interval $t' \to t' + \Delta t'$ and constructing balance equations by summing over all such single event probabilities.† However, in this section an alternate development[24,25] will be discussed to illustrate a complementary point of view. For the sake of clarity, monoenergetic neutrons in an isotropically scattering medium will be considered, and delayed neutrons will be neglected. Energy-dependence and delayed neutrons can be included in the formalism without major modification.[25,26]

† The basic conceptual difference between the "forward" formulation of Section 4.1 and the "backward" formulation of this section is that in the former balance equations for the transition probability $P(N't' \mid Nt)$ are derived by considering changes in the interval t to $t + \Delta t$, whereas changes in the interval t' to $t' + \Delta t$ are considered in deriving balance equations for the latter. The forward and backwards stochastic methods bear much the same relationship to each other as do the flux and adjoint formulations of reactor kinetics.

A neutron will be characterized by its position $\mathbf{r}$ and direction $\boldsymbol{\Omega}$, and its interactions with nuclei of the medium will be described by the total reaction frequency $\Lambda(\mathbf{r}, t)$ plus the probabilities $C_i(\mathbf{r}, t)$ that i neutrons emerge instantaneously from the reaction $(\Sigma_i C_i = 1)$. The stochastic problem is formulated in terms of the probability $p_n(R, t_f; \mathbf{r}, \boldsymbol{\Omega}, t)$ that a neutron at $\mathbf{r}$ with direction $\boldsymbol{\Omega}$ at time t will lead to n neutrons in some region R of $\mathbf{r} - \boldsymbol{\Omega}$ space at time t_f.

An equation for this probability is developed by writing p_n as the product of the probability that a neutron at $\mathbf{r}$ and $\boldsymbol{\Omega}$ has a first reaction times the probability that neutrons emerging from this reaction will lead to n neutrons in R at t_f, plus the probability that the neutron at $\mathbf{r}$ and $\boldsymbol{\Omega}$ does not have a reaction and leads to n neutrons in R at t_f.

The probability that a neutron at $\mathbf{r}$ and $\boldsymbol{\Omega}$ will have a collision in the time interval $t + s \to t + s + \Delta s$ is

$$\Lambda(\mathbf{r} + vs\boldsymbol{\Omega}, t + s)\, \Delta s \exp\left[- \int_0^s \Lambda(\mathbf{r} + s'v\boldsymbol{\Omega}, t + s')\, ds' \right].$$

The probability that i neutrons emerge from the reaction is $C_i(\mathbf{r} + vs\boldsymbol{\Omega}, t + s)$, and the ways in which these emergent neutrons can lead to n neutrons in R at t_f must be considered. If $i = 0$, then the probability is unity for $n = 0$. If $i = 1$, this one neutron can lead to n neutrons (provided $s < t_f - t$); whereas if $i = 2$, one neutron can lead to m neutrons $(m = 0, 1, \ldots, n)$ and the second neutron can lead to $n - m$ neutrons, etc.

$$\begin{aligned} p_n(R, t_f; \mathbf{r}, \boldsymbol{\Omega}, t) &= \int_0^{l(s_B, t_f - t)} ds\, \Lambda(\mathbf{r} + vs\boldsymbol{\Omega}, t + s) \\ &\quad \times \exp\left\{ - \int_0^s \Lambda(\mathbf{r} + vs'\boldsymbol{\Omega}, t + s')\, ds' \right\} \times \Bigg[C_0(\mathbf{r} + vs\boldsymbol{\Omega}, t + s)\, \delta_{n0} \\ &\quad + C_1(\mathbf{r} + vs\boldsymbol{\Omega}, t + s) \int p_n(R, t_f; \mathbf{r} + sv\boldsymbol{\Omega}, \boldsymbol{\Omega}', t + s) \frac{d\boldsymbol{\Omega}'}{4\pi} \\ &\quad + C_2(\mathbf{r} + vs\boldsymbol{\Omega}, t + s) \int \int \sum_{m=0}^{n} p_m(R, t_f; \mathbf{r} + sv\boldsymbol{\Omega}, \boldsymbol{\Omega}', t + s) \\ &\quad \times p_{n-m}(R, t_f; \mathbf{r} + vs\boldsymbol{\Omega}, \boldsymbol{\Omega}'', t + s) \frac{d\boldsymbol{\Omega}'}{4\pi} \frac{d\boldsymbol{\Omega}''}{4\pi} + \cdots \Bigg] + A, \qquad (4.44) \\ &\qquad n = 0, 1, \ldots, \end{aligned}$$

where $l(s_B, t_f - t)$ indicates the lesser of the times s_B required for the neutron to leave the system without a reaction or $t_f - t$. The term A is a correction to account for uncollided neutrons:

$$A = \delta_{n0} A' = \delta_{n0} \exp\left\{ - \int_0^{s_B} \Lambda(\mathbf{r} + vs\boldsymbol{\Omega}, t + s)\, ds \right\}, \qquad t_f - t \geqq s_B,$$

$$A = \delta_{n0} A' = \delta_{n0} \exp\left\{ - \int_0^{t_f - t} \Lambda(\mathbf{r} + vs\boldsymbol{\Omega}, t + s)\, ds \right\}, \qquad t_f - t < s_B$$

and $\mathbf{r} + v(t_f - t)\boldsymbol{\Omega}$ not in R,

$$A = \delta_{n1} A' = \delta_{n1} \exp\left\{ - \int_0^{t_f - t} \Lambda(\mathbf{r} + vs\boldsymbol{\Omega}, t + s)\, ds \right\}, \qquad t_f - t < s_B$$

and $\mathbf{r} + v(t_f - t)\boldsymbol{\Omega}$ in R.

Now introduce the probability generating function

$$G(u \mid \mathbf{r}, \boldsymbol{\Omega}, t) = \sum_{n=0}^{\infty} u^n p_n(R, t_f; \mathbf{r}, \boldsymbol{\Omega}, t). \tag{4.45}$$

An equation for G is obtained by multiplying each of Equations (4.44) by u^n and summing over n:

$$\begin{aligned} &G(u \mid \mathbf{r}, \boldsymbol{\Omega}, t) \\ &\quad = \int_0^{l(s_B, t_f - t)} ds\, \Lambda(\mathbf{r} + vs\boldsymbol{\Omega}, t + s) \\ &\qquad \times \exp\left\{ - \int_0^{s} \Lambda\,(\mathbf{r} + vs'\boldsymbol{\Omega}, t + s')\, ds' \right\} \times \left[C_0(\mathbf{r} + vs\boldsymbol{\Omega}, t + s) \right. \\ &\qquad \left. + \sum_{i=1} C_i(\mathbf{r} + vs\boldsymbol{\Omega}, t + s) \times \left\{ \int G(u \mid \mathbf{r} + vs\boldsymbol{\Omega}, \boldsymbol{\Omega}', t + s) \frac{d\Omega'}{4\pi} \right\}^i \right] + A', \end{aligned} \tag{4.46}$$

unless the third of the defining relations for A applies, in which case the last term is $A'u$.

This integral equation for G may be converted into an integro-differential equation by subtracting Equation (4.46) from the equivalent equation which obtains at $t + \delta s$, and taking the limit $\delta s \to 0$:

$$\begin{aligned} &\boldsymbol{\Omega} \cdot \nabla G(u \mid \mathbf{r}, \boldsymbol{\Omega}, t) + \frac{\partial G}{\partial t}(u \mid \mathbf{r}, \boldsymbol{\Omega}, t) \\ &\quad = \Lambda(\mathbf{r}, t) \times \left\{ G(u \mid \mathbf{r}, \boldsymbol{\Omega}, t) - \sum_{i=0} C_i(\mathbf{r}, t) \left[\int G(u \mid \mathbf{r}, \boldsymbol{\Omega}', t) \frac{d\Omega'}{4\pi} \right]^i \right\}. \end{aligned} \tag{4.47}$$

The probability generating function must satisfy final conditions at time t_f and boundary conditions at $\mathbf{r} = \mathbf{r}_B$ for outward directed, $\mathbf{\Omega} \cdot \hat{n}_B > 0$, neutrons, with $\hat{n}_B$ the outward normal. At $t = t_f$, $p_n = \delta_{n1}$ if the neutron is in R, and $p_n = \delta_{n0}$ otherwise. The appropriate final conditions are

$$G(u \mid \mathbf{r}, \mathbf{\Omega}, t_f) = \begin{cases} u, & \mathbf{r}, \mathbf{\Omega} \in R, \\ 1, & \mathbf{r}, \mathbf{\Omega} \notin R. \end{cases} \tag{4.48}$$

For a neutron that is leaving the system, $p_n = \delta_{n0}$, and the appropriate boundary condition is

$$G(u \mid \mathbf{r}_B, \mathbf{\Omega}, t) = 1, \qquad \mathbf{\Omega} \cdot \hat{n}_B > 0. \tag{4.49}$$

Equations for the moments of the probability distribution can be obtained by differentiating Equation (4.46) or (4.47) with respect to u and evaluating the result for $u = 1$:

$$\begin{aligned} \bar{n}(R, t_f; \mathbf{r}, \mathbf{\Omega}, t) &= \left.\frac{\partial G}{\partial u}\right|_{u=1}, \\ \overline{n(n-1)} &= \left.\frac{\partial^2 G}{\partial u^2}\right|_{u=1}, \quad \text{etc.}, \end{aligned} \tag{4.50}$$

where the overbar indicates a mean value

$$\bar{X} \equiv \sum_{n=0}^{\infty} X_n p_n(R, t_f; \mathbf{r}, \mathbf{\Omega}, t). \tag{4.51}$$

4.9 The Langevin Technique

The theoretical models discussed in Sections 4.1 and 4.8 were based on more or less fundamental stochastic theory in that the developments started with a transition probability and proceeded to the derivation of equations satisfied by these probabilities and their generating functions. Variances, covariances, and correlation functions were interpreted in terms of the product of instantaneous neutron (or precursor) densities in two parts of a reactor, averaged over a hypothetical large ensemble of identical reactors. In a stationary, subcritical reactor the ensemble average could (by the ergodic theory) be replaced by an average over time in a single reactor.

The Langevin technique is based on the assumption that the instantaneous fluctuating neutron density in a stationary reactor satisfies the deterministic kinetics equations when an appropriate fluctuating source term is included in the latter. For a stationary system, the one-group time-dependent neutron diffusion equation is

$$B(\mathbf{r})\bar{N}(\mathbf{r}) = \left(\frac{\partial}{\partial t} - v\nabla\cdot D(\mathbf{r})\,\nabla + v\Sigma_a(\mathbf{r}) - v\nu\Sigma_f(\mathbf{r})\right)\bar{N}(\mathbf{r}) = 0,$$

where $\bar{N}(\mathbf{r})$ is the mean neutron density. The assumption that the instantaneous neutron density $N(\mathbf{r}, t)$ satisfies

$$B(\mathbf{r})N(\mathbf{r}, t) = S(\mathbf{r}, t) \tag{4.52}$$

is the essence of the Langevin method.

Equation (4.52) has a solution

$$N(\mathbf{r}, t) = \int dr'\, dt'\, G(\mathbf{r}', t'\colon \mathbf{r}, t)S(\mathbf{r}', t'), \tag{4.53}$$

where the Green's function satisfies

$$B(\mathbf{r})G(\mathbf{r}', t'\colon \mathbf{r}, t) = \delta(\mathbf{r} - \mathbf{r}')\,\delta(t - t'). \tag{4.54}$$

A quantity of interest is the cross-correlation function of the neutron density at two locations in the reactor. For a stationary reactor, this quantity is defined as

$$\phi_{N_1N_2}(\mathbf{r}_1, \mathbf{r}_2, \tau) = \lim_{T\to\infty} \frac{1}{2T} \int_{-T}^{T} N(\mathbf{r}_1, t)N(\mathbf{r}_2, t + \tau)\, dt. \tag{4.55}$$

In terms of the cross-correlation functions of the fluctuating source,

$$\phi_{S_1S_2}(\mathbf{r}_1, \mathbf{r}_2, \tau) = \lim_{T\to\infty} \frac{1}{2T} \int_{-T}^{T} S(\mathbf{r}_1, t)S(\mathbf{r}_2, t + \tau)\, dt, \tag{4.56}$$

and the formal solution of Equation (4.53), Equation (4.55) becomes†

$$\phi_{N_1N_2}(\mathbf{r}_1, \mathbf{r}_2, \tau) = \int_0^\infty d\tau_1 \int d\mathbf{r}_1'\, G(\mathbf{r}_1', \mathbf{r}_1, \tau_1) \int_0^\infty d\tau_2 \int d\mathbf{r}_2'\, G(\mathbf{r}_2', \mathbf{r}_2, \tau_2) \times \phi_{S_1S_2}(\mathbf{r}_1', \mathbf{r}_2', \tau + \tau_1 - \tau_2). \tag{4.57}$$

In order to evaluate Equation (4.57), it is necessary to compute the Green's functions from Equation (4.54) and to evaluate $\phi_{S_1S_2}$ from

† $G(\mathbf{r}', \mathbf{r}, \tau) \equiv G(\mathbf{r}', t'\colon \mathbf{r}, t' + \tau)$.

Equation (4.56); the latter requires that the fluctuating source S be specified.

Assuming that the fluctuations in the number of reactions of a given type in two nonoverlapping time intervals and/or two nonoverlapping volume elements are statistically independent, the cross-correlation function for the source, $\phi_{S_1S_2}$, can be written

$$\phi_{S_1S_2}(\mathbf{r}_1, \mathbf{r}_2, \tau) = A(\mathbf{r}_1)\, \delta(\mathbf{r}_2 - \mathbf{r}_1)\, \delta(\tau). \tag{4.58}$$

This is frequently referred to as the "white noise" assumption.

The quantity $A(\mathbf{r})$ is the fluctuation source rate spectral density, and may be evaluated from the Schottky formula[20]

$$A(\mathbf{r}) = \sum_r \nu_r{}^2 \bar{l}_r \tag{4.59}$$

where ν_r is the net number of neutrons produced in one reaction of type r and $\bar{l}_r$ is the average reaction rate per cubic centimeter for reactions of type r. The types of reaction processes considered are neutron capture, neutron-fission yielding ν_p prompt neutrons, source neutron emission by an external or spontaneous fission source S_e, leakage out of an incremental volume, and decay of type i delayed neutron precursor. Combining these terms, in the order mentioned, gives

$$\begin{aligned} A(\mathbf{r}) = {} & [v\Sigma_c(\mathbf{r})\bar{N}(\mathbf{r})] + \Big[\sum_{\nu_p}(\nu_p - 1)^2 p(\nu_p) v\Sigma_f(\mathbf{r})\bar{N}(\mathbf{r})\Big] + [S_e(\mathbf{r})] \\ & + [-v\nabla\cdot D(\mathbf{r})\,\nabla\bar{N}(\mathbf{r})] + \Big[\sum_{i=1}^{I} \lambda_i \bar{C}_i(\mathbf{r})\Big]. \end{aligned} \tag{4.60}$$

Because the system is stationary,

$$\sum_{i=1}^{I} \lambda_i \bar{C}_i(\mathbf{r}) = \beta v \bar{\nu} \Sigma_f(\mathbf{r}) \bar{N}(\mathbf{r}),$$

and because

$$\bar{\nu}_p \equiv \sum_{p=0}^{\infty} \nu_p p(\nu_p) \equiv (1 - \beta)\bar{\nu},$$

Equation (4.60) may be written

$$\begin{aligned} A(\mathbf{r}) = {} & [v\Sigma_c(\mathbf{r})\bar{N}(\mathbf{r})] + [\{\overline{(\nu_p - 1)^2} + \beta\bar{\nu}\} v\Sigma_f(\mathbf{r})\bar{N}(\mathbf{r})] + [S_e(\mathbf{r})] \\ & + [-v\nabla\cdot D(\mathbf{r})\,\nabla\bar{N}(\mathbf{r})]. \end{aligned} \tag{4.61}$$

Thus, use of the Langevin technique requires only that the deterministic static equations be solved for $\bar{N}(\mathbf{r})$ and the Green's functions of the deterministic kinetics equations be evaluated. The "white noise" assumption is not appropriate for delayed neutrons, as fluctuations in fission rate are correlated to fluctuations in precursor decay rates at a later time. If the delayed neutron production is considered also as a fluctuating quantity, correction terms on the order of β smaller than the terms just discussed must be introduced.[20]

Although the Langevin technique appears to require several assumptions, it has been shown, by obtaining the statistics of the fluctuating source $S(\mathbf{r}, t)$ from basic stochastic considerations, that the Langevin technique is the equivalent, in the prediction of the statistical properties of the fluctuating neutron density $N(\mathbf{r}, t)$, of more basic theories.[19]

4.10 The Product Density Formulation

The final form of fluctuation theory to be discussed in this chapter is derived by employing the deterministic kinetics equations to compute the probability that a neutron introduced at a certain location in the reactor gives rise to a chain-related descendant neutron in a given elemental volume element in phase space. These probabilities are then employed to calculate product densities, which are product reaction rate probabilities. The product densities are used to determine moments in the stochastic neutron distribution, correlation functions, and other parameters of interest.[14]

Each prompt and delayed neutron that is produced by a fission and the neutron producing the fission are said to be chain related. Delayed neutrons are considered, in this formulation, to be formed at the time of the fission and released later.

Let $W_{S1}(\mathbf{r}, t)\, d\mathbf{r}\, dt$ represent the probability that a source event (e.g., external source, spontaneous fission, (α, n) reaction) occurs within the volume element $d\mathbf{r}$ in the time interval dt. Define $p_S(N_0, N_1, \ldots, N_J)$ as the probability that in a source event N_0 prompt neutrons, N_1 delayed neutrons of type $1, \ldots, N_J$ delayed neutrons of type J are formed. Let $q_{jSR}(\mathbf{r}_0, t_0; \mathbf{r}, \mathbf{v}, t)\, d\mathbf{r}\, d\mathbf{v}\, dt$ represent the probability that if a prompt $(j = 0)$ or delayed $(j \neq 0)$ neutron is formed in a source event at $(\mathbf{r}_0, t_0)$, then this neutron or a chain-related descendant neutron induces a reaction of type R in the phase space interval $d\mathbf{r}\, d\mathbf{v}$ in the time interval dt ($\mathbf{v}$ is the neutron velocity vector).

The first-order product density, $W_{R1}(\mathbf{r}, \mathbf{v}, t)$, which is related to the probability that a reaction of type R takes place within $d\mathbf{r}\, d\mathbf{v}$ about $(\mathbf{r}, \mathbf{v})$ in the time interval dt about t, $W_{R1}(\mathbf{r}, \mathbf{v}, t)\, d\mathbf{r}\, d\mathbf{v}\, dt$, may thus be written

$$W_{R1}(\mathbf{r}, \mathbf{v}, t)\, d\mathbf{r}\, d\mathbf{v}\, dt$$
$$= \int d\mathbf{r}_0 \int_{-\infty}^{t} dt_0\, W_{S1}(\mathbf{r}_0, t_0) \times \sum_{N_0=0}^{\infty} \cdots \sum_{N_J=0}^{\infty} p_S(N_0, \ldots, N_J)$$
$$\times \sum_{j=0}^{J} N_j q_{jSR}(\mathbf{r}_0, t_0; \mathbf{r}, \mathbf{v}, t)\, d\mathbf{r}\, d\mathbf{v}\, dt. \tag{4.62}$$

Employing the definition of the mean number $\bar{\nu}_{jS}$ of prompt ($j=0$) or delayed ($j \neq 0$) neutrons formed in a source event,

$$\bar{\nu}_{jS} \equiv \sum_{N_0=0}^{\infty} \cdots \sum_{N_J=0}^{\infty} N_j p_S(N_0, \ldots, N_J), \tag{4.63}$$

Equation (4.62) becomes

$$W_{R1}(\mathbf{r}, \mathbf{v}, t) = \int d\mathbf{r}_0 \int_{-\infty}^{t} dt_0\, W_{S1}(\mathbf{r}_0, t_0) \sum_{j=0}^{J} \bar{\nu}_{jS} q_{jSR}(\mathbf{r}_0, t_0; \mathbf{r}, \mathbf{v}, t). \tag{4.64}$$

It is assumed that $W_{S1}(\mathbf{r}_0, t_0)$ and $\bar{\nu}_{jS}$ are known. The function $q_{jSR}(\mathbf{r}_0, t_0; \mathbf{r}, \mathbf{v}, t)$ can be computed from the source-free deterministic neutron kinetics equations, under the initial condition that one type-j neutron with a type-j velocity distribution was present at $(\mathbf{r}_0, t_0)$.

The second-order product density $W_{R2}(\mathbf{r}_1, \mathbf{v}_1, t_1; \mathbf{r}_2, \mathbf{v}_2, t_2)$ is defined in terms of the probability $W_{R2}\, d\mathbf{r}_1\, d\mathbf{v}_1\, dt_1\, d\mathbf{r}_2\, d\mathbf{v}_2\, dt_2$ that a reaction of type R is induced within the time interval dt_1 by a neutron in the elemental phase volume $d\mathbf{r}_1\, d\mathbf{v}_1$, and that another reaction of type R is induced within the time interval dt_2 by a neutron in the elemental phase space volume $d\mathbf{r}_2\, d\mathbf{v}_2$. The two neutrons may be members of two distinct fission chains, or members of the same fission chain. When the neutrons are members of two distinct fission chains, the probability that the pair of reactions occurs is $W_{R1}(\mathbf{r}_1, \mathbf{v}_1, t_1) W_{R1}(\mathbf{r}_2, \mathbf{v}_2, t_2)\, d\mathbf{r}_1\, d\mathbf{r}_2\, d\mathbf{v}_1\, d\mathbf{v}_2\, dt_1\, dt_2$. When the two neutrons are chain related, the probability of occurrence of the pair of reactions is compounded of the probabilities that an ancestral source or fission event occurred at an earlier time, that a neutron formed in the event initiated a chain which eventually produced a reaction in $d\mathbf{r}_1\, d\mathbf{v}_1\, dt_1$, and that another neutron formed in the event initiated a chain which eventually produced a reaction in $d\mathbf{r}_2\, d\mathbf{v}_2\, dt_2$.

The probability that a chain-related pair of reactions results from a common ancestral source event is $(t_1 \leqq t_2)$

$$\begin{aligned} &\int d\mathbf{r}_0 \int_{-\infty}^{t} dt_0\, W_{S1}(\mathbf{r}_0, t_0) \sum_{N_0=0}^{\infty} \cdots \sum_{N_J=0}^{\infty} p_S(N_0, \ldots, N_J) \\ &\quad \times \left\{ \sum_{j=0}^{J} N_j q_{jSR}(\mathbf{r}_0, t_0; \mathbf{r}_1, \mathbf{v}_1, t_1)\, d\mathbf{r}_1\, d\mathbf{v}_1\, dt_1 \right. \\ &\quad \times (N_j - 1) q_{jSR}(\mathbf{r}_0, t_0; \mathbf{r}_2, \mathbf{v}_2, t_2)\, d\mathbf{r}_2\, d\mathbf{v}_2\, dt_2 \\ &\quad + \sum_{j=0}^{J} \sum_{\substack{j'=0 \\ j \neq j'}}^{J} N_j q_{jSR}(\mathbf{r}_0, t_0; \mathbf{r}_1, \mathbf{v}_1, t_1)\, d\mathbf{r}_1\, d\mathbf{v}_1\, dt_1 \\ &\quad \left. \times N_{j'} q_{j'SR}(\mathbf{r}_0, t_0; \mathbf{r}_2, \mathbf{v}_2, t_2)\, d\mathbf{r}_2\, d\mathbf{v}_2\, dt_2 \right\}. \end{aligned} \tag{4.65}$$

If the common ancestral event is neutron-induced fission, an analogous expression results. Let $p_F(N_0, \ldots, N_J)$ represent the probability that N_0 prompt neutrons and N_j type-j delayed neutrons $(j = 1, \ldots, J)$ are formed in a neutron-induced fission. Define $q_{jFR}(\mathbf{r}_0, t_0; \mathbf{r}, \mathbf{v}, t)\, d\mathbf{r}\, d\mathbf{v}\, dt$ as the probability that a j-type neutron formed in neutron-induced fission at $(\mathbf{r}_0, t_0)$ initiates a chain which eventually produces a reaction of type R in $d\mathbf{r}\, d\mathbf{v}\, dt$. Then the probability that a neutron-induced fission ancestral event leads to a chain-related reaction pair is given by Equation (4.65), but with the subscript S replaced by F.

Summation of these probabilities for a reaction pair results in $(t_1 \leqq t_2)$

$$\begin{aligned} &W_{R2}(\mathbf{r}_1, \mathbf{v}_1, t_1; \mathbf{r}_2, \mathbf{v}_2, t_2) = W_{R1}(\mathbf{r}_1, \mathbf{v}_1, t_1) W_{R1}(\mathbf{r}_2, \mathbf{v}_2, t_2) \\ &\quad + \int_{-\infty}^{t} dt_0 \int d\mathbf{r}_0 \sum_{j=0}^{J} \sum_{j'=0}^{J} \Big[W_{S1}(\mathbf{r}_0, t_0) \sum_{N_0=0}^{\infty} \cdots \sum_{N_J=0}^{\infty} \{N_j N_{j'} - N_j\, \delta_{jj'}\} \\ &\quad \times p_S(N_0, \ldots, N_J) q_{jSR}(\mathbf{r}_0, t_0; \mathbf{r}_1, \mathbf{v}_1, t_1) \times q_{j'SR}(\mathbf{r}_0, t_0; \mathbf{r}_2, \mathbf{v}_2, t_2) \\ &\quad + W_{F1}(\mathbf{r}_0, t_0) \sum_{N_0=0}^{\infty} \cdots \sum_{N\ =0}^{\infty} \{N_j N_{j'} - N_j\, \delta_{jj'}\} p_F(N_0, \ldots, N_J) \\ &\quad \times q_{jFR}(\mathbf{r}_0, t_0; \mathbf{r}_1, \mathbf{v}_1, t_1) q_{j'FR}(\mathbf{r}_0, t_0; \mathbf{r}_2, \mathbf{v}_2, t_2) \Big]. \end{aligned} \tag{4.66}$$

The quantity W_{F1} can be computed as a special case of Equation (4.64). Definitions analogous to that of Equation (4.63) may be employed to replace the summations over $N_0, \ldots, N_J$ with mean values

$$H_{jj'm} \equiv \sum_{N_0=0} \cdots \sum_{N_J=0} \{N_j N_{j'} - N_j \delta_{jj'}\} p_m(N_0, \ldots, N_J), \qquad m = S, F. \tag{4.67}$$

The mean type-R reaction rate in a finite volume ΔV of space-velocity phase space in the finite-time interval ΔT is given by

$$R_{\Delta V \Delta T} = \int_{\Delta V} d\mathbf{r}\, d\mathbf{v} \int_{\Delta T} dt\, W_{R1}(\mathbf{r}_1, \mathbf{v}_1, t_1), \tag{4.68}$$

and the mean value of the product of the number of reactions in two intervals $\Delta V_1\, \Delta T_1$ and $\Delta V_2\, \Delta T_2$ is

$$\begin{aligned} \overline{R_{\Delta V_1 \Delta T_1} R_{\Delta V_2 \Delta T_2}} &= \int_{\Delta V_1} d\mathbf{r}_1\, d\mathbf{v}_1 \int_{\Delta T_1} dt_1 \int_{\Delta V_2} d\mathbf{r}_2\, d\mathbf{v}_2 \int_{\Delta T_2} dt_2 \\ &\quad \times W_{R2}(\mathbf{r}_1, \mathbf{v}_1, t_1; \mathbf{r}_2, \mathbf{v}_2, t_2) \\ &\quad + \int_{\Delta V_1 \Delta T_1 \cap \Delta V_2 \Delta T_2} d\mathbf{r}_1\, d\mathbf{v}_1 \int dt_1\, W_{R1}(\mathbf{r}_1, \mathbf{v}_1, t_1), \end{aligned} \tag{4.69}$$

where $\Delta V_1\, \Delta T_1 \cap \Delta V_2\, \Delta T_2$ denotes the intersection of $\Delta V_1\, \Delta T_1$ and $\Delta V_2\, \Delta T_2$. Variances and covariances in the type-R reaction rate may be constructed from the solution of Equation (4.69):

$$\operatorname{var}(R_{\Delta V \Delta T}) = \overline{R_{\Delta V \Delta T} R_{\Delta V \Delta T}} - \overline{R_{\Delta V \Delta T}} \times \overline{R_{\Delta V \Delta T}}, \tag{4.70}$$

$$\operatorname{cov}(R_{\Delta V_1 \Delta T_1} R_{\Delta V_2 \Delta T_2}) = \overline{R_{\Delta V_1 \Delta T_1} R_{\Delta V_2 \Delta T_2}} - \overline{R_{\Delta V_1 \Delta T_1}} \times \overline{R_{\Delta V_2 \Delta T_2}}. \tag{4.71}$$

REFERENCES

1. G. E. Hansen, "Assembly of Fissionable Material in the Presence of a Weak Neutron Source," *Nucl. Sci. Eng.* **8,** 709 (1960).
2. H. Hurwitz, D. B. MacMillan, J. H. Smith, and M. L. Storm, "Kinetics of Low Source Reactor Startups, I and II," *Nucl. Sci. Eng.* **15,** 166 and 187 (1963).
3. D. B. MacMillan and M. L. Storm, "Kinetics of Low Source Reactor Startups, III," *Nucl. Sci. Eng.* **16,** 369 (1963).

4. G. I. Bell, "Probability Distribution of Neutrons and Precursors in a Multiplying Assembly," *Ann. Phys.* **21,** 243 (1963).
5. G. I. Bell, W. A. Anderson, D. Galbraith, "Probability Distribution of Neutrons and Precursors in a Multiplying Medium, II," *Nucl. Sci. Eng.* **16,** 118 (1963).
6. D. R. Harris, "Kinetics of Low Source Level," in *Naval Reactors Physics Handbook* (A. Radkowsky, ed.), Vol. I, Section 5.5. TID-7030, USAEC, Washington, D.C., 1964.
7. B. Bars, "Stochastic Fluctuations in the Power Burst of a Reactor," *Nukleonik* **9,** 118 (1967).
8. D. B. MacMillan, "Kinetics of Low Source Reactor Startups—Additional Numerical Results," KAPL-M-6550, Knolls Atomic Power Laboratory (1966).
9. R. P. Feyman, F. deHoffman, and R. Serber, "Dispersion of the Neutron Emission in U-235 Fission," *Nucl. Energy* **3,** 64 (1956).
10. J. D. Orndoff, "Prompt Neutron Periods of Metal Critical Assemblies," *Nucl. Sci. Eng.* **2,** 450 (1957).
11. J. A. Thie, *Reactor Noise*, AEC Monograph. Rowman and Littlefield, New York, 1963.
12. R. E. Uhrig (ed.), *Noise Analysis in Nuclear Systems*, AEC Symposia, TID-7679, USAEC, Washington, D.C., 1964.
13. W. Matthes, "Statistical Fluctuations and Their Correlation in Reactor Neutron Distributions," *Nukleonik* **4,** 213 (1962).
14. D. R. Harris, "Neutron Fluctuations in a Reactor of Finite Size, " *Nucl. Sci. Eng.* **21,** 369 (1965).
15. A. Dalfes, "The Random Processes of a Nuclear Reactor and Their Detection," *Nukleonik* **8,** 94 (1966).
16. A. Dalfes, "Functional Analysis of the Random Processes in Nuclear Reactors," *Nukleonik* **8,** 123 (1967).
17. W. Matthes, "Theory of Fluctuations in Neutron Fields," *Nukleonik* **8,** 87 (1966).
18. M. M. R. Williams, "Reactor Noise in Heterogeneous Systems: I. Plate-Type Elements," *Nucl. Sci. Eng.* **30,** 188 (1967).
19. A. Z. Akcasu and R. K. Osborn, "Application of Langevin's Technique to Space- and Energy-Dependent Noise Analysis," *Nucl. Sci. Eng.* **26,** 13 (1966).
20. J. R. Sheff and R. W. Albrecht, "The Space Dependence of Reactor Noise: I, Theory," *Nucl. Sci. Eng.* **24,** 246 (1966).
21. R. K. Osborn and M. Natelson, "Kinetic Equations for Neutron Distributions," *J. Nucl. Energy, A/B* **19,** 619 (1965).
22. E. D. Courant and P. R. Wallace, "Fluctuations of the Number of Neutrons in a Pile," *Phys. Rev.* **72,** 1038 (1947).
23. W. G. Clark, D. R. Harris, M. Natelson, and J. F. Walter, "Variances and Covariances of Neutron and Precursor Populations in Time Varying Reactors," *Nucl. Sci. Eng.* **31,** 440 (1968).
24. L. Pal, "On the Theory of Stochastic Processes in Nuclear Reactors " *Nuovo Cimento* **7,** 25 (1958).

25. G. I. Bell, "On the Stochastic Theory of Neutron Transport," *Nucl. Sci. Eng.* **21,** 290 (1965).
26. B. F. Zolotar, "Monte Carlo Analysis of Nuclear Reactor Fluctuation Models," *Nucl. Sci. Eng.* **31,** 282 (1968).
27. K. Saito and Y. Taji, "Theory of Branching Processes of Neutrons in a Multiplying Medium," *Nucl. Sci. Eng.* **30,** 54 (1967).
28. W. M. Stacey, Jr., "Stochastic Kinetic Theory for a Space- and Energy-Dependent Zero-Power Nuclear Reactor Model," *Nucl. Sci. Eng.* **36,** 389 (1969).
29. R. K. Osborn and J. M. Nieto, "Detector Effects on the Statistics of Neutron Fluctuations," *Nucl. Sci. Eng.* **26,** 511 (1966).
30. M. Abramowitz and I. A. Stegun, *Handbook of Mathematical Functions.* Dover, New York, 1964.

Chapter 5

XENON SPATIAL OSCILLATIONS

Xenon-135, with a thermal absorption cross section of $\sim 1\text{–}3 \times 10^6$ barns† and a half-life against beta decay of 9.2 hr, is produced by the fission product decay chain

$$
\begin{array}{lcl}
\text{Fission } (^{235}\text{U}) \xrightarrow{6\cdot 1\,\%} {}^{135}\text{Te} \xrightarrow[<1\text{ min}]{\beta^-} & & {}^{135}\text{I} \\
\big\downarrow & & \big\downarrow \beta^- \; 6\cdot 7 \text{ hr} \\
\longrightarrow 0.2\% \longrightarrow & & {}^{135}\text{Xe} \xrightarrow[9\cdot 2\text{ hr}]{\beta^-} {}^{135}\text{Cs}.
\end{array}
$$

The instantaneous production rate of xenon-135 depends upon the iodine-135 concentration, and, hence, upon the local neutron flux history over the past 50 hr or so. On the other hand, the destruction rate of xenon-135 depends upon the instantaneous flux through the neutron absorption process and upon the flux history through the xenon-135 decay process. When the flux is suddenly reduced in a reactor that has been operating at a thermal flux level $>10^{13}$ n/cm^2 sec, the xenon destruction rate decreases dramatically while the xenon production rate is initially unchanged, thus increasing the xenon concentration. The xenon concentration passes through a maximum and decreases to a new equilibrium value as the iodine concentration decays away to a new equilibrium value.

When a flux tilt is introduced into a reactor, the xenon concentration will initially increase in the region in which the flux is reduced, and

† 1 barn $= 10^{-24}$ cm^2.

initially decrease in the region of increased flux for similar reasons. This shift in the xenon distribution is such as to increase (decrease) the multiplication properties of the region in which the flux has increased (decreased), thus enhancing the flux tilt. After a few hours the increased xenon production due to the increasing iodine concentration in the high flux region causes the high flux region to have reduced multiplicative properties, and the multiplicative properties of the low flux region increase due to the decreased xenon production associated with a decreasing iodine concentration. This decreases, and may reverse, the flux tilt. In this manner it is possible, under certain conditions, for the delayed xenon production effects to induce growing oscillations in the spatial flux distribution. Such oscillations are common in the large production reactors at Hanford and Savannah River, and have been observed in smaller reactors such as the Shippingport PWR.

Because of the time scale of the iodine and xenon dynamics, prompt and delayed neutron dynamics may be neglected (i.e., changes in the neutron flux are assumed to occur instantaneously, and the delayed neutron precursors are assumed to be always in equilibrium). Moreover, iodine-135 can be assumed to be formed directly from fission. The appropriate equations are

$$\nabla \cdot D^g(r,t)\,\nabla\phi^g(r,t) - \{\Sigma_a{}^g(r,t) + \Sigma_s{}^g(r,t) + \sigma_x{}^G X(r,t)\,\delta_{g,G}\}\phi^g(r,t)$$
$$+ \sum_{g' \neq g}^{G} \Sigma_s^{g'/g}(r,t)\phi^{g'}(r,t) + \chi_P{}^g \sum_{g'=1}^{G} \nu\Sigma_f^{g'}(r,t)\phi^{g'}(r,t) = 0, \qquad g = 1, \ldots, G \tag{5.1}$$

$$\gamma_i \sum_{g=1}^{G} \Sigma_f{}^g(r,t)\phi^g(r,t) - \lambda_i I(r,t) = \dot{I}(r,t), \tag{5.2}$$

$$\gamma_x \sum_{g=1}^{G} \Sigma_f{}^g(r,t)\phi^g(r,t) + \lambda_i I(r,t) - \lambda_x X(r,t) - \sigma_x{}^G X(r,t)\phi^G(r,t) = \dot{X}(r,t). \tag{5.3}$$

In writing these equations it is assumed that the xenon absorption cross section is zero except in the thermal group ($g = G$). The macroscopic cross sections and diffusion coefficients in Equations (5.1) were defined in Section 1.1. (The absorption cross section does not include xenon.) The quantity $\sigma_x{}^G$ is the microscopic absorption cross section of xenon for thermal neutrons, γ and λ denote yields and decay constants, and I and X are the iodine and xenon concentrations. Changes in the macroscopic cross sections and diffusion coefficients are due to control rod motion or temperature feedback.

The purpose of this chapter is to examine the problem of xenon spatial oscillations. Criteria are developed for predicting stability with respect to these oscillations, the effect of various design variables and the control rod motion upon stability is discussed, and the problem of controlling xenon spatial oscillations is outlined.

5.1 Linear Stability Criteria

One of the features of Equations (5.1)–(5.3) that makes their solution by analytical methods difficult is the nonlinearity introduced by the xenon absorption term (implicit nonlinearities are also introduced by the dependence of the cross sections on the flux via the temperature feedback). Linearizing Equations (5.1)–(5.3) reduces their complexity, but also reduces their applicability to a small region about the equilibrium point. The linearized equations are used principally for investigations of stability; i.e., if a small flux tilt is introduced, will this flux tilt oscillate spatially with an amplitude that diminishes or grows in time?†

The linearized equations are obtained by expanding about the equilibrium point, denoted by a zero subscript,

$$\phi^g(r, t) = \phi_0{}^g(r) + \delta\phi^g(r, t),$$
$$I(r, t) = I_0(r) + \delta I(r, t),$$
$$X(r, t) = X_0(r) + \delta X(r, t),$$

making use of the fact that the equilibrium solutions satisfy the time-independent version of Equations (5.1)–(5.3), and neglecting terms that are nonlinear in $\delta\phi^g$ and δX.

$$\begin{aligned}\nabla \cdot D^g(r)\,\nabla\,\delta\phi^g(r, t) &- \{\Sigma_a{}^g(r) + \Sigma_s{}^g(r) + \sigma_x{}^G(r)X_0(r)\,\delta_{g,G}\}\,\delta\phi^g(r, t)\\ &+ \sum_{g'=g}^{G} \Sigma_s^{g'/g}(r)\,\delta\phi^{g'}(r, t) - \sigma_x{}^G(r)\phi_0{}^G(r)\,\delta X(r, t)\,\delta_{g,G}\\ &+ \chi_P{}^g \sum_{g'=1}^{G} \nu\Sigma_f^{g'}(r)\,\delta\phi^{g'}(r, t) = 0, \qquad (5.4)\\ &\qquad g = 1, \ldots, G,\end{aligned}$$

† The general question of stability is examined in detail in Chapter 6.

$$\gamma_i \sum_{g=1}^{G} \Sigma_f{}^g(r)\, \delta\phi^g(r, t) - \lambda_i\, \delta I(r, t) = \delta \dot{I}(r, t), \tag{5.5}$$

$$\gamma_x \sum_{g=1}^{G} \Sigma_f{}^g(r)\, \delta\phi^g(r, t) + \lambda_i\, \delta I(r, t) - \lambda_x\, \delta X(r, t)$$
$$- \sigma_x{}^G(r) X_0(r)\, \delta\phi^G(r, t) - \sigma_x{}^G(r)\phi_0{}^G(r)\, \delta X(r, t) = \delta \dot{X}(r, t). \tag{5.6}$$

The effect of temperature feedback has been neglected momentarily in writing Equations (5.4)–(5.6), in that the time dependence of the cross sections has been suppressed. Feedback effects will be reintroduced later.

Upon Laplace transforming the time dependence, Equations (5.4)–(5.6) become

$$\nabla \cdot D^g(r)\, \nabla\, \delta\phi^g(r, p) - \{\Sigma_a{}^g(r) + \Sigma_s{}^g(r) + \sigma_x{}^G(r) X_0(r)\, \delta_{g,G}\}\, \delta\phi^g(r, p)$$
$$+ \sum_{g' \neq g}^{G} \Sigma_s^{g'/g}(r)\, \delta\phi^{g'}(r, p) - \sigma_x{}^G(r)\phi_0{}^G(r)\, \delta X(r, p)\, \delta_{g,G}$$
$$+ \chi_P{}^g \sum_{g'=1}^{G} \nu\Sigma_f^{g'}(r)\, \delta\phi^{g'}(r, p) = 0, \tag{5.7}$$
$$g = 1, \ldots, G,$$

$$\gamma_i \sum_{g=1}^{G} \Sigma_f{}^g(r)\, \delta\phi^g(r, p) - (p + \lambda_i)\, \delta I(r, p) = -\, \delta I(r, t = 0), \tag{5.8}$$

$$\gamma_x \sum_{g=1}^{G} \Sigma_f{}^g(r)\, \delta\phi^g(r, p) + \lambda_i\, \delta I(r, p) - (p + \lambda_x + \sigma_x{}^G(r)\phi_0{}^G(r))\, \delta X(r, p)$$
$$- \sigma_x{}^G(r) X_0(r)\, \delta\phi^G(r, p) = -\, \delta X(r, t = 0). \tag{5.9}$$

Equations (5.7)–(5.9) may be written

$$\mathbf{H}\, \delta\mathbf{y} = \delta\mathbf{y}_0, \tag{5.10}$$

where

$$\delta\mathbf{y}(r, p) \equiv \begin{bmatrix} \delta\phi^1(r, p) \\ \vdots \\ \delta\phi^G(r, p) \\ \delta I(r, p) \\ \delta X(r, p) \end{bmatrix}, \qquad \delta\mathbf{y}_0 \equiv \begin{bmatrix} 0 \\ \vdots \\ 0 \\ -\delta I(r, t = 0) \\ -\delta X(r, t = 0) \end{bmatrix},$$

and **H** is composed of the coefficient terms on the left side of Equations (5.7)–(5.9).

The solution of Equation (5.10) is formally

$$\delta \mathbf{y}(r, p) = \mathbf{H}^{-1}(r, p)\, \delta \mathbf{y}_0(r, t = 0). \tag{5.11}$$

Thus, the solutions of Equations (5.7)–(5.9) are related to the initiating perturbations by a transfer function matrix, $\mathbf{H}^{-1}$. The condition that the solutions diminish† in time is equivalent to the condition that the poles of the transfer function (thus the roots of **H**) lie in the left half complex plane. The roots of **H** are the eigenvalues, p, of Equations (5.7)–(5.9), with a homogeneous right-hand side. These homogeneous equations are known as the p-mode equations.

The p-mode equations generally have complex eigenfunctions and eigenvalues, and must be solved numerically except for the simplest geometries. Numerical determination of the p-eigenvalues requires special codes, and has been successful only for slab geometries.

For practical reactor models, it is necessary to resort to approximate methods to evaluate the p-eigenvalues. Two methods that have been successfully employed are the μ-mode and λ-mode approximations.

μ-Mode Approximation The μ-mode approximation[1,2] is motivated by recognition that the only manner in which Equation (5.7) differs from a standard static diffusion theory problem is through the additional term $-\sigma_x{}^G \phi_0{}^G\, \delta X$ in the thermal group balance equation. Using the homogeneous versions of Equations (5.8) and (5.9), this term may be written

$$\sigma_x{}^G(r)\phi_0{}^G(r)\, \delta X(r, p) = N(r, p)\Sigma_f{}^G(r)\, \delta\phi^G(r, p), \tag{5.12}$$

where

$$N(r,p) = \frac{\{1 + \eta(r)\}\{\gamma_x p + \lambda_i(\gamma_x + \gamma_i)\}\, \delta f(r, p) - \eta(r)(\gamma_x + \gamma_i)(p + \lambda_i)f_0(r)}{(1 + (p/\lambda_x) + \eta(r))(p + \lambda_i)(1 + 1/\eta(r))}, \tag{5.13}$$

with

$$\eta(r) \equiv \frac{\sigma_x{}^G(r)\phi_0{}^G(r)}{\lambda_x}, \tag{5.14}$$

† The solutions of Equations (5.7)–(5.9) have an oscillatory time dependence if the roots of **H** have an imaginary component. The requirement that these roots lie in the left half complex plane ensures that these solutions oscillate with a diminishing amplitude.

$$\delta f(r, p) \equiv \sum_{g=1}^{G} \frac{\Sigma_f{}^g(r)}{\Sigma_f{}^G(r)} \cdot \frac{\delta\phi^g(r, p)}{\delta\phi^G(r, p)}, \tag{5.15}$$

$${}_0(r) \equiv \sum_{g=1}^{G} \frac{\Sigma_f{}^g(r)}{\Sigma_f{}^G(r)} \cdot \frac{\phi_0{}^g(r)}{\phi_0{}^G(r)}. \tag{5.16}$$

In applications, the quantity $\delta f(r, p)$ is usually assumed equal to $f_0(r)$.

Using these definitions, the p-mode equations [homogeneous versions of Equations (5.7)–(5.9)] may be written in the equivalent form

$$\begin{aligned} &\nabla \cdot D^g(r) \nabla \delta\phi^g(r, p) - \{\Sigma_a{}^g(r) + \Sigma_s{}^g(r) + \sigma_x{}^G(r) X_0(r)\, \delta_{g,G}\}\, \delta\phi^g(r, p) \\ &\quad + \sum_{g' \neq g}^{G} \Sigma_s^{g'/g}(r)\, \delta\phi^{g'}(r, p) + \chi_P{}^g \sum_{g'=1}^{G} \nu\Sigma_f^{g'}(r)\, \delta\phi^{g'}(r, p) \\ &\quad = N(r, p)\Sigma_f{}^G(r)\, \delta\phi^G(r, p)\, \delta_{g,G}, \\ &\qquad g = 1, \ldots, G. \end{aligned} \tag{5.17}$$

If $N(r, p)$ is real, then the term $N\Sigma_f{}^G$ in Equation (5.17) is formally like a distributed poison, and Equation (5.17) can be solved with standard multigroup diffusion theory codes. In general, $N(r, p)$ is complex because the p-eigenvalues are complex. The essential assumption of the μ-mode approximation is that $N(r, p)$ is real.

There are two types of μ-mode approximations, and they differ in the treatment of the spatial dependence of $N(r, p)$. In the first approximation the spatial dependence is retained explicitly, and $N(r, p)\Sigma_f{}^G(r)$ is treated as a distributed poison, in which case Equations (5.17) become the standard multigroup criticality equations. A value of p is guessed, $N(r, p)$ is evaluated, and Equations (5.17) are solved for the eigenvalue k ($1/k$ multiplies the fission term in the eigenvalue problem). This procedure is repeated until the calculated eigenvalue agrees with the known critical eigenvalue; the corresponding value of p is an approximation to the p-eigenvalue with the largest real part.

An alternate μ-mode approximation (and the one that gives rise to the name μ-mode) results when $N(r, p)$ is assumed to be spatially independent

$$N(r, p) = \mu(p). \tag{5.18}$$

In this case, Equations (5.17) define an eigenvalue problem for the μ-eigenvalues, which can be solved, with a slight modification to the coding, by conventional multigroup diffusion theory codes. To obtain an estimate of the p-eigenvalue from the calculated μ-eigenvalue requires definitions of effective values of $\bar{\eta}$ and $\bar{f}_0$ which account for the spatial dependence of these quantities. In practice, an effective $\bar{\eta}$ is usually defined[1]

$$\bar{\eta} \equiv \frac{\int dr\, \phi_0^{G*}(r)\Sigma_f{}^G(r)\eta(r)\phi_0{}^G(r)}{\int dr\, \phi_0^{G*}(r)\Sigma_f{}^G(r)\phi_0{}^G(r)}, \tag{5.19}$$

an expression which can be motivated by perturbation theory.

Temperature feedback effects are included in the calculation of μ-eigenvalues by perturbation theory.[1,2]

λ-Mode Approximation The λ-mode approximation[2–4] begins with Equations (5.7)–(5.9) and expands the spatial dependence in the eigenfunctions of the neutron balance operator at the equilibrium point (i.e., λ-modes)

$$\begin{aligned} &\nabla \cdot D^g(r)\, \nabla\Psi_n{}^g(r) - \{\Sigma_a{}^g(r) + \Sigma_s{}^g(r) + \sigma_x{}^G(r)X_0(r)\,\delta_{g,G}\}\Psi_n{}^g(r) \\ &\quad + \sum_{g' \neq g}^{G} \Sigma_s^{g'/g}(r)\Psi_n^{g'}(r) + (1/k_n)\chi_P{}^g \sum_{g'=1}^{G} \nu\Sigma_f^{g'}(r)\Psi_n^{g'}(r) = 0, \qquad (5.20) \\ &\qquad\qquad g = 1, \ldots, G, \end{aligned}$$

normalized such that

$$\int dr \left[\sum_{g'=1}^{G} \chi_P^{g'}\Psi_m^{g'*}(r)\right]\left[\sum_{g=1}^{G} \nu\Sigma_f{}^g(r)\Psi_n{}^g(r)\right] = \delta_{m,n}, \tag{5.21}$$

where Ψ_m^{g*} satisfy equations adjoint to Equation (5.20) with appropriate adjoint boundary conditions.

It is convenient to treat thermal feedback explicitly in this approximation by including a power feedback term

$$+\alpha\, \delta f(r)\Sigma_f{}^G(r)\, \delta\phi^G(r, p)\phi_0{}^G(r)$$

on the left side of Equation (5.7), for group G.

When the iodine is eliminated between Equations (5.8) and (5.9), and the flux and xenon are expanded in λ-modes,

$$\delta\phi^g(r, p) = \sum_{n=1}^{N} A_n(p)\Psi_n{}^g(r), \qquad g = 1, \ldots, G, \tag{5.22}$$

$$\delta X(r, p) = \sum_{n=1}^{N} B_n(p)\Sigma_f{}^G(r)\Psi_n{}^G(r), \qquad g = 1, \ldots, G, \tag{5.23}$$

the biorthogonality relation of Equation (5.21) may be used to reduce Equations (5.7)–(5.9) to a set of $2N$ algebraic equations in the unknowns A_n and B_n, with inhomogeneous terms involving spatial integrals containing $\delta X(r, t=0)$ and $\delta I(r, t=0)$. These equations may be written as a transfer function relation between the inhomogeneous terms $\mathbf{R}$ and the column vector $\mathbf{A}(p)$ containing the A_n and B_n.

$$\mathbf{A}(p) = \hat{\mathbf{H}}(p) \cdot \mathbf{R}. \tag{5.24}$$

Again, the condition for stability is that the poles of $\hat{\mathbf{H}}$ lie in the left half complex p-plane. When $N = 1$ in the expansion of Equations (5.22) and (5.23), Equation (5.24) may be reduced to the scalar relation[4]

$$A_1(p) = \hat{H}'(p) \cdot R' \tag{5.25}$$

where

$$\hat{H}'(p) = [(p - p_1)(p - p_2)]^{-1} \tag{5.26}$$

and

$$\begin{aligned} p_1 &= -p_r + i(c - p_r^{\,2})^{1/2}, \\ p_2 &= -p_r - i(c - p_r^{\,2})^{1/2}, \end{aligned} \tag{5.27}$$

with

$$p_r = \frac{\lambda_x}{2}\left\{\left(1 + \frac{\lambda_i}{\lambda_x} + \eta\right) - \frac{\eta}{\Omega}\left[\frac{(\gamma_i + \gamma_x)\eta}{1 + \beta} - \gamma_x\right]\right\},$$

$$c = \lambda_i \lambda_x \left\{(1 + \eta) + \frac{\eta(\gamma_i + \gamma_x)}{\Omega}\left[1 - \frac{\eta}{1 + \beta}\right]\right\}.$$

The parameters η, Ω, and β, which characterize the reactor in this formulation, are defined as

$$\eta \equiv \frac{1}{\lambda_x} \frac{\int dr\, \Psi_1^{G*}(r) \sigma_x{}^G(r) \Sigma_f{}^G(r) \phi_0{}^G(r) \Psi_1{}^G(r)}{\int dr\, \Psi_1^{G*}(r) \Sigma_f{}^G(r) \Psi_1{}^G(r)}, \tag{5.28}$$

$$\Omega \equiv \frac{1/k_1 - 1/k_0}{\delta f \int dr\, \Psi_1^{G*}(r) \Sigma_f{}^G(r) \Psi_1{}^G(r)} - \frac{\int dr\, \Psi_1^{G*}(r) \alpha(r) \Sigma_f{}^G(r) \phi_0{}^G(r) \Psi_1{}^G(r)}{\int dr\, \Psi_1^{G*}(r) \Sigma_f{}^G(r) \Psi_1{}^G(r)}, \tag{5.29}$$

$$\beta \equiv \frac{\delta f(\gamma_i + \gamma_x) \int dr\, \Psi_1^{G*}(r) \Sigma_f{}^G(r) \phi_0{}^G(r) \Psi_1{}^G(r)}{\int dr\, \Psi_1^{G*}(r) \lambda_x X_0(r) \Psi_1{}^G(r)} - 1. \tag{5.30}$$

The quantity δf was defined previously as the ratio of the total fission rate to the thermal group fission rate, and an effective spatially independent value has been assumed. The fundamental and first harmonic λ-eigenvalues are denoted by k_0 and k_1, respectively.

The requirement that the poles of $\hat{H}'(p)$ lie in the left half complex p-plane (i.e., that $p_r > 0$) defines a relationship among η, Ω, and β. In practice, $\beta \cong \eta$ has been found to be a good approximation, so that the stability requirement defines a curve in the η-Ω phase plane, as shown in Figure 5.1.

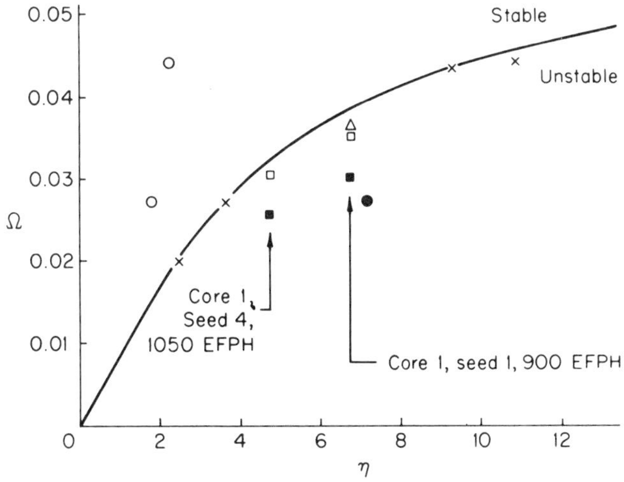

FIGURE 5.1. λ-mode linear xenon stability criterion. KEY: *PWR results*: □, calculated with feedback; ■, calculated, no feedback; △, inferred from experiment. *Calculated transients*: ○, decaying oscillation; ×, neutral oscillation; ●, growing oscillation.

The effect of physical parameters upon xenon spatial stability can be traced through Equations (5.28) and (5.29) and Figure 5.1. The quantity Ω is primarily determined by the eigenvalue separation $1/k_1 - 1/k_0$. A reactor becomes less stable when the eigenvalue separation decreases, which occurs when the dimensions are increased, when the migration length is decreased, or when the power distribution is flattened. A negative power coefficient ($\alpha < 0$) increases Ω, thus making a reactor more stable. The quantity η is proportional to the thermal flux level, $\phi_0{}^G$. An increase in thermal flux level is generally destabilizing (increasing η), but may be stabilizing if $\alpha < 0$ (increasing Ω); i.e., for $\alpha < 0$, an increase in thermal flux moves the point characterizing a given reactor in Figure 5.1 to the right and up. It is interesting that an increase in thermal flux level can, under some circumstances, be stabilizing, although this is not generally the case.

If $\Omega > \gamma_i$, Figure 5.1 predicts stability independent of the value of η. Physically, Ω is a measure of the "reactivity" required to excite the first harmonic λ-mode in the presence of power feedback, and γ_i is a measure of the maximum "reactivity" that can be introduced by iodine decay into xenon.

The parameters η and Ω can be evaluated using standard multigroup diffusion theory codes. A fundamental λ-mode flux and first harmonic flux and adjoint calculations are required. The integrals in Equations (5.28) and (5.29) may be performed with any code that computes perturbation theory type integrals. Computation of first harmonic flux and adjoint requires either that the problem is symmetric so that zero flux boundary conditions may be located on node lines or that the Wielandt iteration scheme be employed.

Several comparisons with experiment and numerical simulation[5] are indicated in Figure 5.1. The location of the symbol indicates the prediction of the stability criterion, and the type symbol indicates the experimental or numerical result.

At 900 effective full power hours (EFPH), Core 1 Seed 1 of the Shippingport reactor experienced planar xenon oscillations with a doubling time of 30 hours. Using this doubling time and the calculated value for η, an experimental Ω may be inferred that agrees with the calculated Ω to within 3%. Core 1 Seed 4 of the Shippingport reactor was observed to be quite unstable at 893 EFPH, and to be slightly unstable at 1397 EFPH. These observations are consistent with the predictions of the stability criterion at 1050 EFPH.

The finite difference approximations to Equations (5.1)–(5.3) were solved numerically for a variety of two-dimensional, three-group reactor models. These same reactor models were evaluated for stability with the λ-mode stability criterion. The results depicted in Figure 5.1 indicate that the predictions of the stability criterion were generally reliable.

In the analysis of this section the total power was assumed to be held constant, and the effects of nonlinearities and control rod motion on the stability were neglected. Although the effects of xenon dynamics upon the total power in an uncontrolled reactor can be evaluated,[6] most reactors can be controlled to yield a constant power output. The treatment of nonlinearities and control rod motion are discussed in the next two sections.

5.2 Nonlinear Stability Criterion

The extended methods of Lyapunov, which are discussed in Section 6.4, are applied to derive a stability criterion[7] which includes the nonlinear terms that were neglected in the previous section.

Employing a one-group neutronics model and retaining the prompt neutron dynamics, and expanding the flux, iodine, and xenon about their equilibrium states, the equations governing the reactor dynamics may be written in matrix format as

$$\dot{\mathbf{y}}(r,t) = \mathbf{L}(r)\mathbf{y}(r,t) + \mathbf{g}(r,t), \tag{5.31}$$

where

$$\mathbf{y}(r,t) \equiv \begin{bmatrix} \delta\phi(r,t) \\ \delta X(r,t) \\ \delta I(r,t) \end{bmatrix}, \qquad \mathbf{g}(r,t) \equiv -\begin{bmatrix} v\sigma_x\,\delta X\,\delta\phi + \alpha v\Sigma_f(\delta\phi)^2 \\ \sigma_x\,\delta X\,\delta\phi \\ 0 \end{bmatrix},$$

$$\mathbf{L}(r) \equiv \begin{bmatrix} (v\nabla\cdot D\nabla - v\Sigma_a + v\nu\Sigma_f - v\sigma_x X_0 - \alpha v\Sigma_f\phi_0) & -v\sigma_x\phi_0 & 0 \\ (\gamma_x\Sigma_f - \sigma_x X_0) & -(\lambda_x + \sigma_x\phi_0) & \lambda_i \\ \gamma_i\Sigma_f & 0 & -\lambda_i \end{bmatrix}. \tag{5.32}$$

where v is the neutron speed, α is the power feedback coefficient, and the other notation has been defined previously.

A Lyapunov functional may be chosen as

$$V[\mathbf{y}] = \tfrac{1}{2}\int_R dr\,\mathbf{y}^T(r,t)\mathbf{y}(r,t). \tag{5.33}$$

The condition for stability (asymptotic stability) in the sense of Lyapunov is that $\dot{V}$ evaluated along the system trajectory defined by Equation (5.31) is negative semidefinite (definite).

$$
\begin{aligned}
\dot{V} &= \tfrac{1}{2}\int_R dr\,[\dot{\mathbf{y}}^T\mathbf{y} + \mathbf{y}^T\dot{\mathbf{y}}] \\
&= \tfrac{1}{2}\int_R dr\,\mathbf{y}^T(\mathbf{L}^* + \mathbf{L})\mathbf{y} + \int_R dr\,\mathbf{g}^T\mathbf{y} \\
&\leqq -\mu\int_R dr\,\mathbf{y}^T\mathbf{y} + \left(\int_R dr\,\mathbf{g}^T\mathbf{g}\right)^{1/2}\left(\int_R dr\,\mathbf{y}^T\mathbf{y}\right)^{1/2},
\end{aligned}
\tag{5.34}
$$

where μ is the smallest eigenvalue of

$$
\tfrac{1}{2}(\mathbf{L}^* + \mathbf{L})\boldsymbol{\varphi}_n = -\mu_n\boldsymbol{\varphi}_n. \tag{5.35}
$$

Thus, the condition for stability is

$$
\mu \geqq \frac{\left(\int_R dr\,\mathbf{g}^T\mathbf{g}\right)^{1/2}}{\left(\int_R dr\,\mathbf{y}^T\mathbf{y}\right)^{1/2}}. \tag{5.36}
$$

For a given reactor model and equilibrium state, characterized by μ, relation (5.36) defines the domain of perturbations for which a stable response will be obtained. For asymptotic stability, the inequality must obtain in relation (5.36).

The linear eigenvalue problem, Equation (5.35), which must be solved for μ, involves the matrix $\mathbf{L}$ of Equation (5.32) and its Hermitean adjoint $\mathbf{L}^*$. The matrix operator $\tfrac{1}{2}(\mathbf{L}^* + \mathbf{L})$ is self-adjoint with a spectrum of real eigenvalues and a complete set of orthogonal eigenfunctions.

The foregoing choice of Lyapunov functional is not unique. As a consequence, this type of analysis provides sufficient, but not necessary, conditions for stability.

5.3 Control-Induced Xenon Spatial Transient Phenomena

Inclusion of the control system in a stability analysis is difficult primarily because of the difficulty encountered in analytically representing the motion of discrete control rods required to maintain criticality. Control rod motion has a profound effect upon the transient response to a perturbation in the equilibrium state in many cases, however, and neglect of this effect may completely invalidate the stability analysis. In this section, some results of a study in which the one-group neutron, xenon, and iodine equations were solved numerically are discussed.

A radially and axially zoned cylindrical model, with properties and dimensions typical of a large pressurized water reactor, is depicted in Figure 5.2. Flow of coolant upward through the core was represented explicitly in the calculation, and thermal feedback effects were incorporated into the neutronics model. Although radial power shifts occurred in response to control rod motion, the transients were essentially axial in character.

Control rod motion required to maintain criticality dominated the transient response. Two transients initiated by first building up steady-state xenon (thermal flux level $\sim 2 \times 10^{13}$ n/cm^2 sec), then shutting the reactor down from 48 to 60 hr, then returning to full power, were simulated. In one case criticality was maintained by the motion of a discrete bank of rods located in region C of Figure 5.2, and in the other case a uniform poison was varied to maintain criticality. The relative power in edit regions XX and YY during the two transients is shown in Figure 5.3. The transient response is quite stable when poison is employed, but is barely stable when discrete rods are used. This emphasizes the fact, when the control rod motion interacts with the mode of xenon oscillation,† that it is necessary to represent explicitly the control rod motion in assessing the stability of a reactor.

The transient response characteristics depend upon the initiating perturbation. Two transients were simulated in which steady-state xenon was built up, then the reactor was shut down for 2 or 12 hr, then returned to full power. Rods in Region C were moved to maintain criticality. Representative local xenon concentrations are depicted in Figure 5.4. (The solid lines are for the 12-hr shutdown, and the dashed lines are for the 2-hr shutdown.) The deviation of these xenon concentrations at 50 or 60 hr, from the values at 48 hr, is a measure of the initiating perturbation for the subsequent transient. The corresponding power transients are depicted in Figure 5.5. The transient ensuing from the 2-hr shutdown is quite stable, while the 12-hr shutdown produces a transient that is barely stable. This difference in transient response is due not only to the difference in initiating perturbation, but also to the difference in control rod motion required to compensate the initiating perturbation for the two cases.

Thermal feedback influences axial xenon transients through several interrelated phenomena. Negative moderator feedback limits local power

† In some cases this may not occur. For instance, the axial motion of an azimuthally symmetric bank of control rods would have no interaction with azimuthal xenon oscillations.

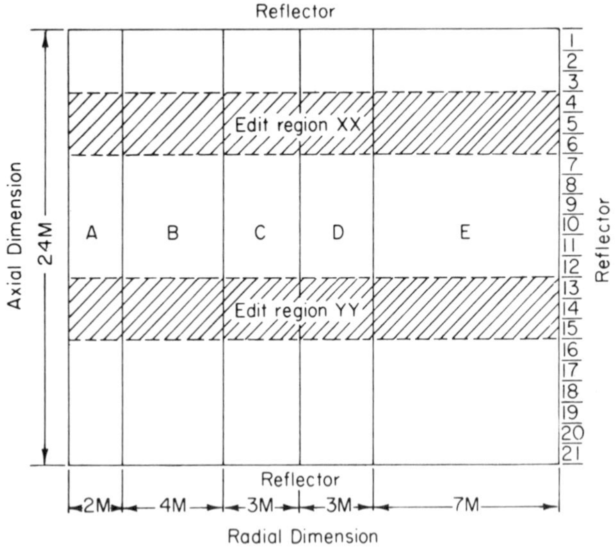

FIGURE 5.2. Cylindrical reactor model. (M = migration length.)

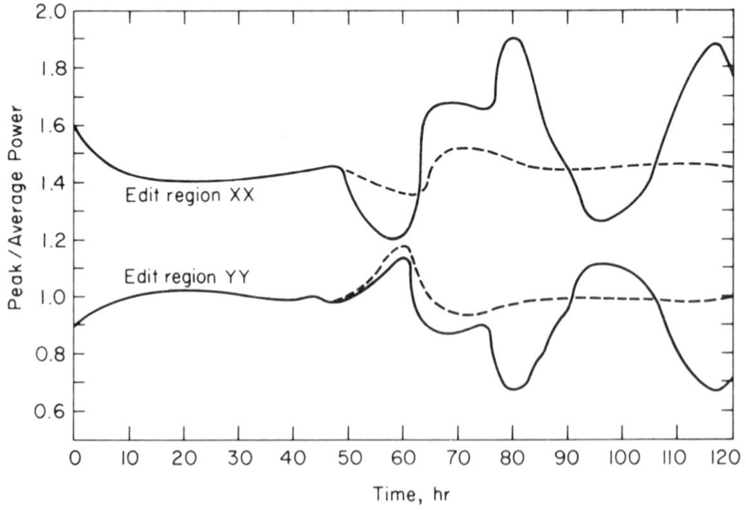

FIGURE 5.3. Influence of control rod motion on xenon-power spatial oscillations. KEY: —, discrete rods; - - -, uniform poison.

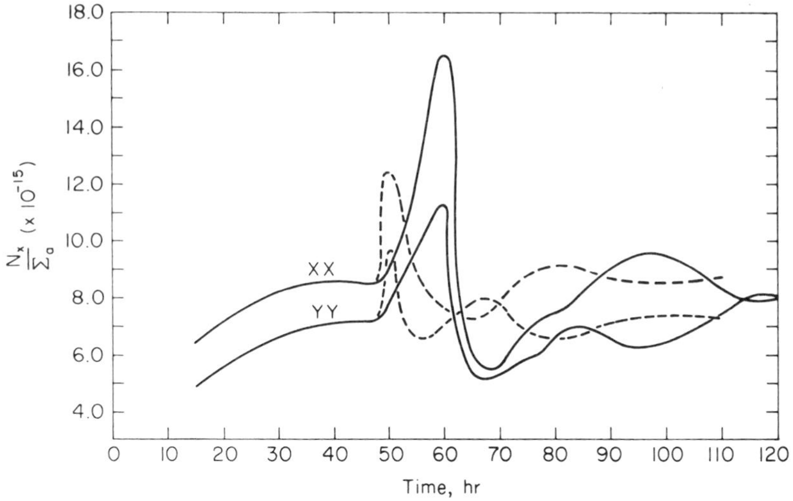

FIGURE 5.4. Variation in local xenon concentrations. KEY: XX, region A, 21 M from bottom; YY, region A, 8 M from bottom.

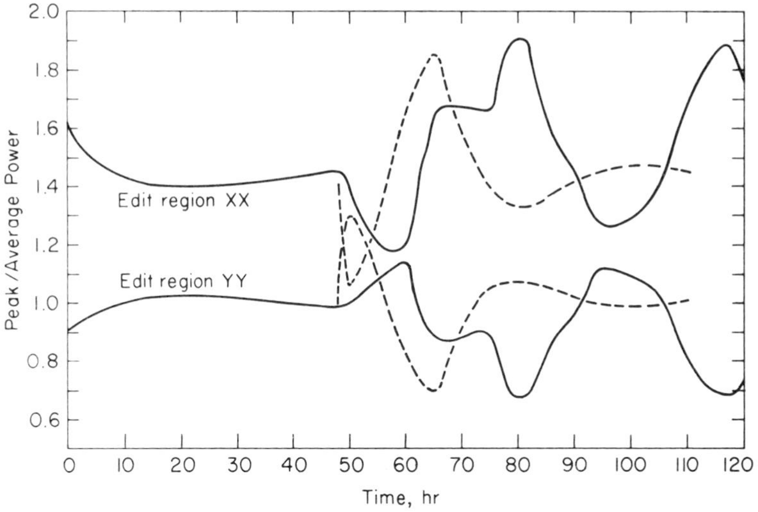

Figure 5.5. Influence of initiating perturbation upon characteristics of xenon-power spatial oscillations. KEY: —, 12-hr shutdown; - - -, 2-hr shutdown.

peaking by increasing the local leakage. Negative fuel feedback limits local power peaking by decreasing the local multiplication properties. Negative feedback limits local power tilting and thus is stabilizing. However, introducing design changes which increase the magnitude of negative feedback coefficients may have an indirect destabilizing effect by flattening the axial power distribution, thus decreasing the axial eigenvalue separation and thereby making the reactor more susceptible to axial oscillations. Opposite arguments can be made for positive feedback coefficients.

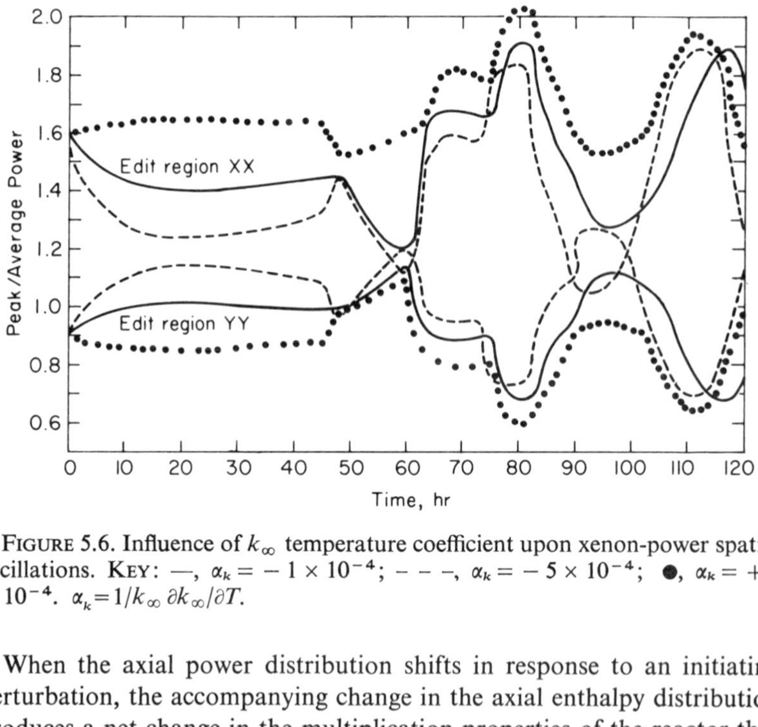

FIGURE 5.6. Influence of k_∞ temperature coefficient upon xenon-power spatial oscillations. KEY: —, $\alpha_k = -1 \times 10^{-4}$; – – –, $\alpha_k = -5 \times 10^{-4}$; ●, $\alpha_k = +5 \times 10^{-4}$. $\alpha_k = 1/k_\infty \, \partial k_\infty/\partial T$.

When the axial power distribution shifts in response to an initiating perturbation, the accompanying change in the axial enthalpy distribution produces a net change in the multiplication properties of the reactor that must be compensated by a change in control rod position. Depending upon whether the compensating rod motion tends to enhance or diminish the original power shift, this effect can be destabilizing or stabilizing, respectively. This effect is destabilizing for negative feedback coefficients when a downward rod motion produces an upward power shift, but is stabilizing when a downward rod motion produces a downward power shift.† Again, an opposite argument can be made for positive feedback coefficients.

Three identical 12-hr shutdown transients, differing only in the value of the fuel feedback coefficients (α_k), were simulated. For these transients, the rods in region C were in the bottom part of the core and a downward rod motion produced an upward power shift. Thus, negative feedback produced a destabilizing control rod feedback effect. A negative moderator feedback coefficient was used in all cases. The relative power in edit regions XX and YY is shown for the three transients in Figure 5.6. The effect of varying the fuel feedback coefficient on the equilibrium axial power distribution is apparent at 48 hr. A large negative fuel feedback coefficient led to diverging oscillations, whereas converging oscillations resulted when a large positive fuel feedback coefficient was used.

An identical set of transients was simulated, but with the critical eigenvalue biased so that the rods in region C were in the top part of the core, where a downward rod motion produced a downward power shift. In this case, negative feedback produced a stabilizing control rod feedback effect. Diverging oscillations resulted when a large positive fuel feedback coefficient was used, whereas a large negative fuel feedback coefficient led to converging oscillations.

5.4 Control of Xenon-Power Spatial Oscillations

In controlling large thermal power reactors, the control rod motion necessary to maintain criticality during a change in operating conditions may, as illustrated in the previous section, induce transients in the spatial xenon and power distributions. Owing to the delayed effects associated with the xenon production, a control action that seems appropriate to suppress an instantaneous power tilt may ultimately aggravate power tilting. Thus, in those reactors which are susceptible to xenon-power spatial oscillations, the eventual, as well as the immediate, consequences of any control action must be considered.

For cores with extensive in-core instrumentation and on-line flux monitoring and display equipment, experienced operators have been successful in suppressing incipient oscillations by appropriate control action. In general, the philosophy has been to control so as to return the

† If the rods move in from the top of the core and the coolant flows upward.

xenon concentration to its equilibrium distribution. For example, if the flux in a region increases, passes through a maximum, and starts to decrease towards its equilibrium value, the decrease in flux may signal an increasing xenon concentration and the appropriate control actions would be such as to increase the flux in the region to inhibit this xenon buildup. Another school of thought favors controlling the flux distribution with sufficient precision so that a significant deviation from the equilibrium distribution for more than a few minutes is not allowed.

The general problem of controlling xenon-power spatial oscillations is, in principle, amenable to analysis by the techniques of modern control theory. However, the application of this theory to xenon spatial oscillations has been limited.[8,9] The large number of possible states complicates the problem to an incredible extent, even for relatively simple reactor models.

In this section the ideas of dynamic programming are used to develop a computing algorithm for the determination of the optimal control sequence from among a set of allowable control actions.[8] Employing a one-group neutronics model, the iodine and xenon dynamics are described by

$$\dot{I}(r,t) = \gamma_i \Sigma_f(r,t)\phi(r,t) - \lambda_i I(r,t), \tag{5.37}$$

and

$$\dot{X}(r,t) = \gamma_x \Sigma_f(r,t)\phi(r,t) + \lambda_i I(r,t) - [\lambda_x + \sigma_x \phi(r,t)]\,X(r,t). \tag{5.38}$$

For the time scale of interest in such problems, changes in the neutron flux and temperature distributions may be assumed to occur instantaneously, and delayed neutrons may be assumed in equilibrium. The neutron flux satisfies the equation,

$$\begin{aligned}\nabla \cdot D(r,t)\,\nabla\phi(r,t) + \nu\Sigma_f(r,t)\phi(r,t) - \Sigma_a(r,t)\phi(r,t)&\\ - \sigma_x X(r,t)\phi(r,t) - R(r,t)\phi(r,t) = 0&\end{aligned} \tag{5.39}$$

and the temperature distribution (T) satisfies a functional equation which is represented as

$$H[T, \phi] = 0. \tag{5.40}$$

In the previous equations, R is the macroscopic control rod cross section. The time dependence of the diffusion coefficient (D), the fission cross section (Σ_f), and the absorption cross section (Σ_a) enters via a change in temperature.

The physical limitations on the control rod displacements constrain the allowable control action to a certain domain Π, a fact which is represented by the relation

$$R(r, t) \in \Pi. \tag{5.41}$$

In order to make the notation more compact, the following definitions are employed:

$$\begin{aligned}
F &\equiv \Sigma_f \phi, \\
\mathbf{y} &\equiv \begin{pmatrix} I \\ X \end{pmatrix}, \\
\gamma &\equiv \begin{pmatrix} \gamma_i \\ \gamma_x \end{pmatrix}, \\
\mathbf{A} &\equiv \begin{pmatrix} -\lambda_i & 0 \\ \lambda_i & -\left(\lambda_x + \dfrac{\sigma_x}{\Sigma_f} F\right) \end{pmatrix}.
\end{aligned} \tag{5.42}$$

With these definitions, Equations (5.37)–(5.47) become

$$\begin{aligned}
\dot{\mathbf{y}}(r, t) &= \mathbf{A}(r, F(r, \mathbf{y}(r, t), R(r, t)))\mathbf{y}(r, t) + \gamma F(r, \mathbf{y}(r, t), R(r, t)) \\
&\equiv \mathbf{f}(\mathbf{y}(r, t), F(r, y(r, t), R(r, t))),
\end{aligned} \tag{5.43}$$

$$R(r, t) \in \Omega, \tag{5.44}$$

$$\Phi[\phi, T] = 0. \tag{5.45}$$

Relation (5.44) implies that the control (R) satisfies relations (5.39) and (5.41); i.e., is within the domain of physically realizable and allowed control actions, and is such as to maintain criticality. Equation (5.45) implies that the neutron and temperature balance equations, Equations (5.39) and (5.40), are satisfied.

The general objective of a control program may be to minimize power peaking or xenon buildup, or some combination of these two. Thus, an optimality functional (i.e., a quantity whose minimization is the object of the control program) can be written in the form†

$$J[R] = \int_{t_0}^{t_f} dt\, G(\mathbf{y}(t), F(\mathbf{y}(t), R(t)). \tag{5.46}$$

(The spatial argument has been suppressed.)

† The optimality functional is written as an integral over time of the quantity G, which is itself an integral over space.

The problem of determining the optimal control of xenon-power spatial transients can now be formulated succinctly: Find the control action (R) that minimizes the functional of Equation (5.46) subject to the constraints of Equations (5.43)–(5.45).

Suppose the interval $t_0 \leqq t \leqq t_f$ is divided into N segments $t_{n-1} \leqq t \leqq t_n$, the state of the system at the initial time of each interval (t_{n-1}) is denoted by $\mathbf{y}_{n-1}$, and the type of control action during each interval is denoted by R_n. Further, denote the value of the functional of Equation (5.46) evaluated over the interval $t_{n-1} \leqq t \leqq t_n$ by J_n. The value of J_1 depends upon the initial state $\mathbf{y}_0$ and the type of control action taken during the interval $t_0 \leqq t \leqq t_1$, this control being denoted by R_1. The value of J_2 depends upon $\mathbf{y}_1$ and R_2, but $\mathbf{y}_1$ depends upon $\mathbf{y}_0$ and R_1. In general, the value of the functional evaluated over the interval $t_{n-1} \leqq t \leqq t_n$, J_n, depends on R_n and $\mathbf{y}_{n-1}$, but the latter depends upon $\mathbf{y}_0$, $R_1, \ldots, R_{n-1}$. Physically, the power peaking and xenon concentrations, and, hence, the value of J_n, depend upon the type of control action taken in the interval $t_{n-1} \leqq t \leqq t_n$ and the xenon and iodine concentrations at t_{n-1}, but the latter depend upon the initial xenon and iodine concentrations and the type of control actions taken during the time $t_0 \leqq t \leqq t_{n-1}$. Thus, the optimal control problem amounts to solving the equation

$$S[\mathbf{y}_0] = \min_{R_1, \ldots, R_N \in \Omega} \left\{ \sum_{n=1}^{N} J_n[\mathbf{y}_0, R_1, \ldots, R_n] \right\}. \tag{5.47}$$

Equation (5.47) calls for the simultaneous minimization of the functional within the braces { } with respect to the N control actions $R_1, \ldots, R_N$, which is an overwhelming task for a practical reactor model. This task is reduced to a manageable magnitude by using the ideas of dynamic programming.

The only term in the sum of Equation (5.47) that depends upon R_N is J_N. Thus, Equation (5.47) may be written

$$S[\mathbf{y}_0] = \min_{R_1, \ldots, R_{N-1} \in \Omega} \left\{ \sum_{n=1}^{N-1} J_n[\mathbf{y}_0, R_1, \ldots, R_n] + \min_{R_N \in \Omega} J_N[\mathbf{y}_0, R_1, \ldots, R_N] \right\}. \tag{5.48}$$

The minimization of J_N with respect to R_N is carried out for all possible values of $\mathbf{y}_0$, $R_1, \ldots, R_{N-1}$, and the result is denoted by

$$S_N[\mathbf{y}_0, R_1, \ldots, R_{N-1}] \equiv S_N[\mathbf{y}_{N-1}] \equiv \min_{R_N \in \Omega} J_N[\mathbf{y}_0, R_1, \ldots, R_N]. \quad (5.49)$$

Using Equation (5.49), Equation (5.48) becomes

$$S[\mathbf{y}_0] = \min_{R_1, \ldots, R_{N-1} \in \Omega} \left\{ \sum_{n=1}^{N-1} J_n[\mathbf{y}_0, R_1, \ldots, R_n] + S_N[\mathbf{y}_0, R_1, \ldots, R_{N-1}] \right\}. \quad (5.50)$$

In Equation (5.48) or (5.50), the only terms which depend upon R_{N-1} are J_{N-1} and S_N, which leads to the equivalent equation,

$$S[\mathbf{y}_0] = \min_{R_1, \ldots, R_{N-2} \in \Omega} \left\{ \sum_{n=1}^{N-2} J_n[\mathbf{y}_0, R_1, \ldots, R_n] + \min_{R_{N-1} \in \Omega} \{ J_{N-1}[\mathbf{y}_0, R_1, \ldots, R_{N-1}] + S_N[\mathbf{y}_0, R_1, \ldots, R_{N-1}] \} \right\}. \quad (5.51)$$

The minimization of $J_{N-1} + S_N$ with respect to R_{N-1} is carried out for all possible values of $\mathbf{y}_0$, $R_1, \ldots, R_{N-2}$, and the result is denoted by

$$S_{N-1}[\mathbf{y}_0, R_1, \ldots, R_{N-2}] \equiv S_{N-1}[\mathbf{y}_{N-2}] \equiv \min_{R_{N-1} \in \Omega} \{J_{N-1} + S_N\}, \quad (5.52)$$

so that Equation (5.51) may be written

$$S[\mathbf{y}_0] = \min_{R_1, \ldots, R_{N-2} \in \Omega} \left\{ \sum_{n=1}^{N-2} J_n[\mathbf{y}_0, R_1, \ldots, R_n] + S_{N-1}[\mathbf{y}_0, R_1, \ldots, R_{N-2}] \right\}. \quad (5.53)$$

This procedure may be continued consecutively until the initial interval is reached, as only J_k, $k \geqq n$ depend upon R_n. This is a consequence of the fact that a control action can affect the present and future states of the system, but cannot affect the past states. The appropriate generalization of Equation (5.52) is

$$S_{N-k}[\mathbf{y}_0, R_1, \ldots, R_{N-k-1}] = \min_{R_{N-k} \in \Omega} \{ J_{N-k}[\mathbf{y}_0, R_1, \ldots, R_{N-k}] + S_{N-k+1}[\mathbf{y}_0, R_1, \ldots, R_{N-k}] \}, \quad (5.54)$$

or, equivalently,

$$S_{N-k}[\mathbf{y}_{N-k-1}] = \min_{R_{N-k} \in \Omega} \{J_{N-k}[\mathbf{y}_{N-k-1}, R_{N-k}] + S_{N-k+1}[\mathbf{y}_{N-k-1}, R_{N-k}]\} \tag{5.55}$$

The value of $S_1[\mathbf{y}_0]$ which is obtained from the sequence of minimizations indicated by Equation (5.54) for $k = 1, \ldots, N-1$, plus the minimization of Equation (5.49), is identical to the value $S[\mathbf{y}_0]$ defined by Equation (5.47). The problem of simultaneously minimizing a functional with respect to N types of control action [Equation (5.47)] has thus been replaced by the problem of minimizing N functionals each with respect to a single control action [Equations (5.49) and (5.54)]. These latter minimizations must each be performed for many values of the variables $\mathbf{y}_0, R_1, \ldots, R_{N-k}$, however. Equations (5.49) and (5.54) embody the dynamic programming approach to the optimization problem of Equation (5.47). This formalism has been applied to the reactor model described in the previous section.[8]

REFERENCES

1. T. R. England, G. L. Hartfield, and R. K. Deremer, "Xenon Spatial Stability in Large Seed Blanket Reactors," WAPD-TM-606, Bettis Atomic Power Laboratory (1966).
2. S. Kaplan, "Modal Analysis of Xenon Stability," in *Naval Reactors Physics Handbook*, A. Radkowsky (ed.), Vol. I, pp. 977–1010. TID-7030, USAEC, Washington, D.C., 1964.
3. D. Randall and D. S. St. John, "Xenon Spatial Oscillations," *Nucleonics* **16,** 82 (1958).
4. W. M. Stacey, Jr., "Linear Analysis of Xenon Spatial Oscillations," *Nucl. Sci. Eng.* **30,** 448 (1967).
5. R. J. Hooper, R. A. Rydin, and W. M. Stacey, Jr., "Verification of a Xenon Spatial Stability Criterion," *Nucl. Sci. Eng.* **34,** 344 (1968).
6. J. Chernick, "The Dynamics of a Xenon Controlled Reactor," *Nucl. Sci. Eng.* **8,** 233 (1960).
7. W. M. Stacey, Jr., "A Non-Linear Xenon Stability Criterion for a Spatially Dependent Reactor Model," *Nucl. Sci. Eng.* **35,** 395 (1969).
8. W. M. Stacey, Jr., "Optimal Control of Xenon-Power Spatial Transients," *Nucl. Sci. Eng.* **33,**162 (1968).
9. D. M. Wiberg, "Optimal Feedback Control of Spatial Xenon Oscillations in a Nuclear Reactor," Ph.D. thesis, California Institute of Technology (1965).

Chapter 6

STABILITY

In a nuclear reactor operating at steady-state conditions, an equilibrium obtains among the interacting neutronic, thermodynamic, hydrodynamic, thermal, xenon, etc., phenomena. The state of the reactor is defined in terms of the values of the state functions† associated with each of these phenomena (e.g., the neutron flux, the coolant enthalpy, the coolant pressure). If a reactor is perturbed from an equilibrium state, will the ensuing state (1) remain bounded within some specified domain of the state functions, (2) return to the equilibrium state after a sufficiently long time, or (3) diverge from the equilibrium state in that one or more of the state functions takes on a shape outside a specified domain of state functions? This is the question of stability.

A great deal of effort has been expended in developing mathematical models describing the state variables which are appropriate for different reactor types, and in performing stability analyses on models that incorporated space-independent (point) neutron kinetics. Much of this work is summarized in References 1–3. Relatively little work has been done on space-dependent reactor stability.

† In a spatially dependent system, such as a nuclear reactor, the state of the system is defined in terms of spatially dependent state functions. When the spatial dependence is discretized by one of the approximations of Chapter 1, the state of the system is defined in terms of discrete state variables. When dealing with state functions, it is necessary to work in a generalized function space, whereas the discrete state variables can be considered in an N-dimensional Euclidean phase space.

This chapter outlines a theory appropriate for the stability analysis of spatially dependent reactor models. The first three sections are concerned with the stability analysis of the coupled system of ordinary differential equations that results when the spatial dependence of Equations (1.1) and (1.2), and of the corresponding equations for the other state functions, are approximated by one of the methods of Chapter 1. The final section outlines the extended Lyapunov theory for the stability analysis of the coupled partial differential equations [Equations (1.1) and (1.2) and the corresponding equations for the other state functions] which describe spatially continuous systems.

6.1 Classical Linear Stability Analysis

The finite-difference, time-synthesis, nodal, or point kinetics approximations of Chapter 1, and the corresponding approximations to the other state function equations, may be written as a coupled set of ordinary differential equations relating the discrete state variables y_i,

$$\dot{y}_i(t) = f_i(y_1(t), \ldots, y_N(t)), \qquad i = 1, \ldots, N, \tag{6.1}$$

where, for instance, y_i may be the neutron flux at node i and y_{I+j} may be the coolant enthalpy at node j. The coupling among the equations arises because the cross sections in the neutronics equations depend upon the local temperature, density, and xenon concentration, because the temperature, density, and xenon concentration depend upon the local flux, and because neutron- and heat-diffusion and coolant transport introduces a coupling among the value of the state variables at different locations.

Equation (6.1) may be written as a vector equation,

$$\dot{\mathbf{y}}(t) = \mathbf{f}(\mathbf{y}(t)), \tag{6.2}$$

where the components of the column vectors $\mathbf{y}$ and $\mathbf{f}$ are the y_i and f_i, respectively. The equilibrium state $\mathbf{y}_e$ satisfies

$$\mathbf{f}(\mathbf{y}_e) = 0. \tag{6.3}$$

If the solution of Equation (6.2) is expanded about $\mathbf{y}_e$,

$$\mathbf{y}(t) = \mathbf{y}_e + \hat{\mathbf{y}}(t), \tag{6.4}$$

and the part of the right-hand side of Equation (6.2) which is linear in $\hat{\mathbf{y}}$ is separated out, then Equation (6.2) may be written

$$\dot{\hat{\mathbf{y}}}(t) = \mathbf{h}(\mathbf{y}_e)\hat{\mathbf{y}}(t) + \mathbf{g}(\mathbf{y}_e, \hat{\mathbf{y}}(t)). \tag{6.5}$$

The matrix $\mathbf{h}$ has constant elements, some of which may depend upon the equilibrium state.

Classical linear stability analysis proceeds by ignoring the nonlinear term $\mathbf{g}$ in Equation (6.5). It is readily shown that the condition for the stability of the linearized equations is that the real part of all eigenvalues of the matrix $\mathbf{h}$ are negative. To illustrate this, apply a permutation transformation which diagonalizes $\mathbf{h}$ to the linear approximation to Equation (6.5):

$$\mathbf{P}^T\dot{\hat{\mathbf{y}}}(t)\mathbf{P} = \mathbf{P}^T\mathbf{h}\mathbf{P}\mathbf{P}^T\hat{\mathbf{y}}(t)\mathbf{P},$$

since

$$\mathbf{P}^T\mathbf{P} = \mathbf{P}\mathbf{P}^T = \mathbf{I}.$$

Define $\mathbf{X}(t) \equiv \mathbf{P}^T\hat{\mathbf{y}}(t)\mathbf{P}$. Then the transformed equations are decoupled

$$\dot{X}_i(t) = \omega_i X_i(t), \qquad i = 1, \ldots, N$$

where ω_i are the eigenvalues of $\mathbf{h}$. The solutions of these equations subject to $X_i(0) = X_{i0}$ are

$$X_i(t) = X_{i0}e^{\omega_i t}, \qquad i = 1, \ldots, N,$$

which may be written in vector notation as

$$\mathbf{X}(t) = \mathbf{\Gamma}(t)\mathbf{X}_{i0}$$

where $\mathbf{\Gamma}(t) = \operatorname{diag}(\exp(\omega_i t))$. Hence,

$$\hat{\mathbf{y}}(t) = \mathbf{P}\mathbf{X}(t)\mathbf{P}^T = \mathbf{P}\mathbf{\Gamma}(t)\mathbf{X}_{i0}\mathbf{P}^T.$$

If $\operatorname{Re}(\omega_i) < 0$, then $\lim_{t\to\infty}\hat{\mathbf{y}}(t) = 0$; i.e., the state of the system returns to the equilibrium state. If $\operatorname{Re}(\omega_i) > 0$, one or more of the components of $\hat{\mathbf{y}}$ approach ∞ as $t \to \infty$, and the system is unstable.

Thus, stability analysis of the linearized equations amounts to determining if the eigenvalues of the $\mathbf{h}$ matrix are in the left- (stable) or right- (unstable) half complex plane. This determination may be accomplished most readily by Laplace transforming the linearized equation into the frequency domain and then applying one of the methods of linear control theory[4] (e.g., Bode, Nyquist, root locus, Hurwitz) that have been developed explicitly for this purpose.

The xenon stability criteria of Section 5.1 are examples of this type of analysis.

6.2 Lyapunov's Method

The method of Lyapunov attempts to draw certain conclusions about the stability of the solution of Equation (6.5) without any knowledge of this solution. Essential to this method is the choice of a scalar function $V(\hat{\mathbf{y}})$ which is a measure of a metric distance of the state $\mathbf{y} = \mathbf{y}_e + \hat{\mathbf{y}}$ from the equilibrium state $\mathbf{y}_e$. Let $\hat{\mathbf{y}}(t, \hat{\mathbf{y}}_0)$ be the solution of Equation (6.5) for the initial condition $\hat{\mathbf{y}}(t=0) = \hat{\mathbf{y}}_0$. If it can be shown that $V(\hat{\mathbf{y}}(t, \hat{\mathbf{y}}_0))$ will be small when $V(\hat{\mathbf{y}}_0)$ is small, then $\mathbf{y}_e$ is a stable equilibrium state. If, in addition, it can be shown that $V(\hat{\mathbf{y}}(t, \hat{\mathbf{y}}_0))$ approaches zero for large times, then $\mathbf{y}_e$ is an asymptotically stable equilibrium state.

Define a scalar function $V(\hat{\mathbf{y}})$ that depends upon all the state variables $\hat{y}_i$ and which has the following properties in some region R about the equilibrium state $\mathbf{y}_e$:

(1) $V(\hat{\mathbf{y}})$ is positive definite; i.e., $V(\hat{\mathbf{y}}) > 0$ if $\hat{\mathbf{y}} \neq 0$, $V(\hat{\mathbf{y}}) = 0$ if $\hat{\mathbf{y}} = 0$.

(2) $\lim_{\hat{\mathbf{y}} \to 0} V(\hat{\mathbf{y}}) = 0$, $\lim_{\hat{\mathbf{y}} \to \infty} V(\hat{\mathbf{y}}) = \infty$.

(3) $V(\hat{\mathbf{y}})$ is continuous in all its partial derivatives; i.e., $\partial V/\partial y_i$ exist and are continuous for $i = 1, \ldots, N$.

(4) $\dot{V}(\hat{\mathbf{y}})$ evaluated along the solution of Equation (6.5) is nonpositive; i.e.,

$$\dot{V}(\hat{\mathbf{y}}) = \sum_{i=1}^{N} \frac{\partial V}{\partial \hat{y}_i} \dot{\hat{y}}_i = \sum_{i=1}^{N} \frac{\partial V}{\partial \hat{y}_i} f_i \leqq 0.$$

A scalar function $V(\hat{\mathbf{y}})$ satisfying properties (1)–(4) is a Lyapunov function.

Three theorems based on the Lyapunov function can be stated about the equilibrium solution of Equation (6.5).

1. *Stability Theorem.* If a Lyapunov function exists in some region R about $\mathbf{y}_e$, then this equilibrium state is stable for all initial perturbations in R; i.e., for all initial perturbations $\hat{\mathbf{y}}_0$ in R, the solution of Equation (6.5), $\hat{\mathbf{y}}(t, \hat{\mathbf{y}}_0)$, remains within the region R for all $t > 0$.

2. *Asymptotic Stability Theorem.* If a Lyapunov function exists in some region R about $\mathbf{y}_e$, and in addition $\dot{V}$ evaluated along the solution of Equation (6.5) is negative definite ($\dot{V} < 0$ if $\hat{\mathbf{y}} \neq 0$, $\dot{V} = 0$ if $\hat{\mathbf{y}} = 0$) in R, then this equilibrium state is asymptotically stable for all initial perturbations in R; i.e., for all initial perturbations $\hat{\mathbf{y}}_0$ in R, the solution of Equation (6.5) is $\hat{\mathbf{y}}(t, \hat{\mathbf{y}}_0) = 0$ after a sufficiently long time.

3. *Instability Theorem.* If a scalar function $V(\hat{\mathbf{y}})$ which has properties (1)–(3) exists in a region R, and $\dot{V}$ evaluated along the solution of Equation (6.5) does not have a definite sign, then the equilibrium state $\mathbf{y}_e$ is unstable for initial perturbations in R; i.e., for initial perturbations $\hat{\mathbf{y}}_0$ in R, the solution of Equation (6.5), $\hat{\mathbf{y}}(t, \hat{\mathbf{y}}_0)$, does not remain in R for all $t > 0$.

Mathematical proofs of these theorems can be constructed.† Rather than repeat these proofs, it is informative to consider a topological argument. Properties (1)–(3) define a concave upward surface (the function V) in the phase space defined by the $\hat{y}_i$. This surface has a minimum within the region R at $\hat{y}_1 =, \ldots, = \hat{y}_N = 0$ by property (1), and increases monotonically in value as the $\hat{y}_i$ increase, by properties (2) and (3). Thus, contours can be drawn in the hyperplane of the $\hat{y}_i$ representing the locus of points at which V has a given value. These contours are concentric about the equilibrium state $\hat{y}_i = 0$, $i = 1, \ldots, N$. Proceeding outward from this origin, the value of V associated with each contour is greater than the value associated with the previous contour.

The outward normal to these contours is

$$\sum_{i=1}^{N} \frac{\partial V}{\partial \hat{y}_i} \hat{\imath},$$

where $\hat{\imath}$ denotes the unit vector in the direction in phase space associated with the state variable $\hat{y}_i$. The "direction" in which the state of the system is moving in phase space is given by

$$\sum_{i=1}^{N} \dot{\hat{y}}_i \hat{\imath} = \sum_{i=1}^{N} f_i \hat{\imath}.$$

† Many excellent discussions of Lyapunov's methods for the study of the stability of dynamic systems are available in the literature. References 5–7 contain extensive developments and proofs.

For stability, the "direction" in which the state of the system is moving must never be toward regions in which V is larger (i.e., never away from the equilibrium state)

$$\left(\sum_{i=1}^{N} \frac{\partial V}{\partial \hat{y}_i} \hat{\imath}\right) \cdot \left(\sum_{j=1}^{N} f_j \hat{\jmath}\right) = \sum_{i=1}^{N} \frac{\partial V}{\partial \hat{y}_i} f_i \leqq 0.$$

For asymptotic stability, the state of the system must always move toward regions in which V is smaller (i.e., always move toward the equilibrium state). Thus, the inequality must always obtain in the foregoing relation. If the system can move away from the equilibrium state into regions of larger V, then the $\leqq$ is replaced by $>$ in the foregoing relation, and the equilibrium state is unstable.

The Lyapunov method yields the same results obtained in the previous section in the limit in which the nonlinear terms are small. The function

$$V(\hat{\mathbf{y}}) = \hat{\mathbf{y}}^T \hat{\mathbf{y}} = \sum_{i=1}^{N} (\hat{y}_i)^2$$

satisfies properties (1)–(3). Making use of Equation (6.5),

$$\dot{V}(\hat{\mathbf{y}}) = \sum_{i=1}^{N} \frac{\partial V}{\partial \hat{y}_i} \dot{\hat{y}}_i = 2 \sum_{i=1}^{N} \hat{y}_i \dot{\hat{y}}_i = 2\hat{\mathbf{y}}^T \dot{\hat{\mathbf{y}}},$$

$$\dot{V}(\hat{\mathbf{y}}) = 2\{\hat{\mathbf{y}}^T \mathbf{h} \hat{\mathbf{y}} + \hat{\mathbf{y}}^T \mathbf{g}\}.$$

If the region R is defined such that

$$|\hat{\mathbf{y}}^T \mathbf{h} \hat{\mathbf{y}}| > |\hat{\mathbf{y}}^T \mathbf{g}|,$$

a sufficient condition for $\dot{V}$ to be negative definite in R is that $\hat{\mathbf{y}}^T \mathbf{h} \hat{\mathbf{y}}$ is negative definite, a sufficient condition for which is that the eigenvalues of $\mathbf{h}$ have negative real parts. This is the same result obtained in the linear analysis of the previous section. In this case, the Lyapunov method provides, in addition, the region R within which the linear analysis is valid.

In applying the Lyapunov method, construction of a suitable Lyapunov function is the main consideration. Because the Lyapunov function for a system of equations is not unique, the analysis yields sufficient, but not necessary, conditions for stability.

To date, there has been relatively little application of Lyapunov's methods in the stability analysis of spatially dependent reactor models that are approximated by coupled ordinary differential equations. The method, however, has been applied successfully to the point reactor model.[3, 8]

6.3 Input–Output/Computer Simulation

When the necessary accuracy of representation requires the set of equations (6.1) to be large in number, stability analysis by the methods previously discussed may become too cumbersome to be practical. In such cases it may be simpler to solve Equations (6.1) on a digital computer and to infer stability properties from these solutions. The solution of Equations (6.1), for any one of the state variables, may be written

$$y_i(t) = F[X(t-\tau)], \qquad \tau \leqq t, \tag{6.6}$$

where F is a functional which relates a known (input) variation in one of the state variables X to the state variable y_i. For a given $X(t-\tau)$, $y_i(t)$ can be obtained numerically as the computer solution to Equations (6.1). This information can be used to determine the impulse response function for the linear approximation to the system, and thus the system transfer function.

Equation (6.6) may be written as a functional expansion

$$\begin{aligned} y_i(t) &= \int_0^\infty h_1(\tau_1)X(t-\tau_1)\,d\tau_1 \\ &\quad + \int_0^\infty \int_0^\infty h_2(\tau_1, \tau_2)X(t-\tau_1)X(t-\tau_2)\,d\tau_1\,d\tau_2 + \cdots, \end{aligned} \tag{6.7}$$

where h_1 is the impulse function for the linear approximation, and h_n, $n > 1$, are higher order kernels. The transfer function is related to h_1 by

$$H(s) = \int_0^\infty h_1(\tau)\exp(-s\tau)\,d\tau. \tag{6.8}$$

The impulse response function can be obtained by cross-correlating the calculated y_i (output) with the X (input). The cross-correlation function is defined

$$\phi_{xy}(\tau) \equiv \frac{1}{T}\int_0^T X(t-\tau)y_i(t)\,dt. \tag{6.9}$$

Using Equation (6.7), Equation (6.8) becomes

$$\begin{aligned} \phi_{xy}(\tau) &= \frac{1}{T}\int_0^T dt \int_0^\infty d\tau_1\, h_1(\tau_1)X(t-\tau_1)X(t-\tau) \\ &\quad + \frac{1}{T}\int_0^T dt \int_0^\infty d\tau_1 \int_0^\infty d\tau_2\, h_2(\tau_1, \tau_2)X(t-\tau_1)X(t-\tau_2)X(t-\tau) \\ &\quad + \cdots. \end{aligned} \tag{6.10}$$

If the reactor properties are time invariant and not supercritical, then the integration over t may be displaced by τ_1 without changing the value of the integral. Then Equation (6.10) may be written

$$\phi_{xy}(\tau) = \int_0^\infty d\tau_1\, h_1(\tau_1)\phi_{xx}(\tau - \tau_1) + \int_0^\infty d\tau_1 \int_0^\infty d\tau_2\, h_2(\tau_1, \tau_2)\phi_{xxx}(\tau_2 - \tau_1, \tau - \tau_1) + \int_0^\infty d\tau_1 \int_0^\infty d\tau_2 \int_0^\infty d\tau_3\, h_3(\tau_1, \tau_2, \tau_3) \times \phi_{xxxx}(\tau_2 - \tau_1, \tau_3 - \tau_1, \tau - \tau_1) + \cdots \tag{6.11}$$

where autocorrelation functions of various orders are defined:

$$\phi_{xx}(\tau_1) \equiv \frac{1}{T}\int_0^T X(t - \tau_1)X(t)\,dt,$$

$$\phi_{xxx}(\tau_1, \tau_2) \equiv \frac{1}{T}\int_0^T X(t - \tau_1)X(t - \tau_2)X(t)\,dt, \qquad \text{etc.} \tag{6.12}$$

Periodic pseudorandom ternary signals have a number of useful properties which facilitate obtaining information about h_1 from Equation (6.11).

(1) The signal takes on one of three discrete values $+1, 0, -1$.

(2) The signal is discontinuous and may change its value at uniformly spaced time increments Δt.

(3) The signal is periodic, with period $T = (3^n - 1)\Delta t$, where n is an integer. The sequence repeats with all values multiplied by -1 every half period, thus having a mean value of zero over a period.

(4) The jth bit of the signal, C_j, is generated from the linear recursion relation

$$C_j = \sum_{i=1}^{n} a_i C_{j-i} \tag{6.13}$$

where the right-hand side is reduced to modulo 3 to obtain allowed values. Recursion coefficients are given by Hooper and Gyftopoulos.[9]

(5) The autocorrelation functions have the following properties:

(a) $\phi_{xx}(\tau) \simeq \dfrac{2T^2}{3^{n+1}} \sum_{m=-\infty}^{\infty} (-1)^m\, \delta\left(\tau - \dfrac{mT}{2}\right).$

(b) All odd-order autocorrelation functions (ϕ_{xxx}, ϕ_{xxxxx}, etc.) are identically zero.

If the settling time of the system is short in comparison with the period of the signal (i.e., $h_1(\tau) \simeq 0$ for $\tau > T/2$), use of property (5) and Equation (6.9) in Equation (6.11) yields

$$h_1(\tau) \simeq \frac{1}{2(\Delta t)^2 - 3^{n-1}} \int_0^T X(t-\tau) y_i(t)\, dt, \qquad 0 < \tau < T/2. \quad (6.14)$$

To ensure that this condition is met in a practical sense, theoretical limits have been derived for choosing T and Δt such that frequency components in the range $\omega_{\min} \leqq \omega \leqq \omega_{\max}$ will be determined with no more than 3 decibels of attenuation.[10] These relations are

$$\Delta t \leqq \frac{2.07}{\omega_{\max}}, \qquad T \geqq \frac{3.51}{\omega_{\min}}. \quad (6.15)$$

This method has been applied successfully to the stability analysis of a relatively complicated reactor model that represented the neutronics and core thermal-hydraulics by a nodal approximation.[11]

6.4 Lyapunov's Methods for Distributed Parameter Systems

In the previous sections of this chapter, methods were discussed for analyzing the coupled set of ordinary differential equations that result when the spatial dependence of Equations (1.1) and (1.2) and the corresponding equations for the other state variables are approximated by one of the procedures outlined in Chapter 1. A more basic characterization of a reactor system is in terms of spatially distributed state functions, rather than discrete state variables. These state functions satisfy coupled partial differential equations [e.g., Equations (1.1) for neutrons, the heat conduction equation], which may be written

$$\dot{y}_i(r,t) = f_i(y_1(r,t), \ldots, y_N(r,t), r), \qquad i = 1, \ldots, N, \quad (6.16)$$

where y_i is a state function (e.g., neutron group flux) and f_i denotes a spatially dependent operation involving scalers and spatial derivatives upon the state functions. These equations may be written

$$\dot{\mathbf{y}}(r, t) = \mathbf{f}(\mathbf{y}(r, t), r) \tag{6.17}$$

where $\mathbf{y}$ is a column vector of the y_i, and $\mathbf{f}$ is a column vector of the operations denoted by the f_i.

The extension of Lyapunov's methods to systems described by state functions involves the choice of a functional that provides a measure of the distance of the vector of state functions $\mathbf{y}$ from a specified equilibrium state, $\mathbf{y}_{eq}$. The distance between two states $\mathbf{y}_a$ and $\mathbf{y}_b$, $d[\mathbf{y}_a, \mathbf{y}_b]$, is defined as the metric on the product state function space consisting of all possible functions of position that the component state functions can take on.†

An equilibrium state $\mathbf{y}_{eq}(r)$ satisfying

$$\mathbf{f}(\mathbf{y}_{eq}(r), r) = 0$$

is stable if, for any number $\varepsilon > 0$, it is possible to find a number $\delta > 0$ such that when

$$d[\mathbf{y}_0(r), \mathbf{y}_{eq}(r)] < \delta,$$

then

$$d[\mathbf{y}(r, t; \mathbf{y}_0), \mathbf{y}_{eq}(r)] < \varepsilon \quad \text{for} \quad t \geqq 0,$$

where $\mathbf{y}(r, t; \mathbf{y}_0)$ is the solution of Equation (6.17) with the initial condition $\mathbf{y}(r, 0) = \mathbf{y}_0(r)$. If in the limit of large t the distance $d[\mathbf{y}(r, t; \mathbf{y}_0), \mathbf{y}_{eq}]$ approaches zero, then $\mathbf{y}_{eq}$ is asymptotically stable.

Stability Theorem.[12] For an equilibrium state $\mathbf{y}_{eq}(r)$ to be stable, it is necessary and sufficient that in some neighborhood of $\mathbf{y}_{eq}(r)$ that includes the equilibrium state there exists a functional $V[\mathbf{y}]$ with the following properties.

(1) V is positive-definite with respect to $d[\mathbf{y}, \mathbf{y}_{eq}]$; i.e., for any $C_1 > 0$, there exists a $C_2 > 0$ depending on C_1 such that when $d[\mathbf{y}, \mathbf{y}_{eq}] > C_1$, then $V[\mathbf{y}] > C_2$ for all $t \geqq 0$, and

$$\lim_{d[\mathbf{y}, \mathbf{y}_{eq}] \to 0} V[\mathbf{y}] = 0.$$

† The function space consisting of all possible functions of position that the state function y_i can take on is a component function space, and the product of such component function spaces for the different state functions is the state function space on which the vector functions $\mathbf{y}$ are defined.

(2) V is continuous with respect to $d[\mathbf{y}, \mathbf{y}_{eq}]$; i.e., for any real $\varepsilon > 0$, there exists a real $\delta > 0$ such that $V[\mathbf{y}] < \varepsilon$ for all $\mathbf{y}$ in the state function space for $0 < t < \infty$, when $d[\mathbf{y}_0, \mathbf{y}_{eq}] < \delta$.

(3) $V[\mathbf{y}]$ evaluated along any solution $\mathbf{y}$ of Equation (6.17) is non-increasing in time for all $t > 0$ provided that $d[\mathbf{y}_0, \mathbf{y}_{eq}] < \delta_0$, where δ_0 is a sufficiently small positive number.

Asymptotic Stability Theorem.[12] If, in addition to these three conditions, $V[\mathbf{y}]$ evaluated along any solution to Equation (6.17) approaches zero for large t, then the equilibrium state is asymptotically stable.

The same type of topological arguments made in support of the theorems of Section 6.2 are appropriate here, if the state space of that section is generalized to a state function space. Detailed analytical proofs of these theorems may be found in the literature.[7,12]

Construction of a suitable Lyapunov functional is the essential aspect of applying the theory of this section. Experience in this regard is quite limited, although some simple examples are worked out by Hsu,[12] and one example is given in Section 5.2. Although the conditions cited in the theorems are necessary and sufficient for stability, the V-functional chosen may result in more restrictive stability criteria than would be obtained from another V-functional. Thus, stability analyses employing Lyapunov functionals yield only sufficient conditions for stability.

Any of the other techniques used in mathematics for the stability analysis of coupled partial differential equations may be appropriate for the stability analysis of spatially dependent nuclear reactor models. One of these, the method of comparison functions, has been developed and used to analyze idealized reactor models.[13,14]

REFERENCES

1. T. J. Thompson and J. G. Beckerly (eds.), *The Technology of Nuclear Reactor Safety*, Vol. 1. M.I.T. Press, Cambridge, Massachusetts (1964).
2. "Proc. Conf. Transfer Function Measurement and Reactor Stability Analysis," ANL-6205, Argonne National Laboratory (1960).
3. L. Shotkin, "Mathematical Methods in Reactor Dynamics." Academic Press, New York (in preparation).
4. H. Chestnut, *Systems Engineering Tools*, Section 5.4. Wiley, New York, 1965.
5. J. Lasalle and S. Lefschetz, *Stability by Lyapunov's Direct Method.* Academic Press, New York, 1961.

6. N. N. Krasovskii, *Stability of Motion.* Stanford University Press, Stanford, California, 1963.
7. V. I. Zubov, *Methods of A. M. Lyapunov and Their Application*, USAEC translation AEC-tr-4439. USAEC Div. Tech. Info., Washington, D.C., 1961.
8. E. P. Gyftopoulos, "Theoretical and Experimental Criteria for Nonlinear Reactor Stability," *Nucl. Sci. Eng.* **26,** 26 (1966).
9. R. J. Hooper and E. P. Gyftopoulos, "On the Measurement of Characteristic Kernels of a Class of Nonlinear Systems," in *Neutron Noise, Waves, and Pulse Propagation*, pp. 343–356. AEC Symposium Series, No. 9 (Conf. 660206) Gainesville, Florida, 1966.
10. J. D. Balcomb, "A Cross-Correlation Method for Measuring the Impulse Response of Reactor Systems," Ph.D. thesis, Massachusetts Institute of Technology (1961).
11. R. A. Rydin, "The Application of Ternary Signals to the Numerical Evaluation of Complicated Nonlinear Dynamic System Models," KAPL-M-6946, Knolls Atomic Power Laboratory (1968).
12. C. Hsu, "Control and Stability Analysis of Spatially Dependent Nuclear Reactor Systems," ANL-7322, Argonne National Laboratory (1967).
13. W. E. Kastenberg and P. L. Chambre, "On the Stability of Nonlinear Space-Dependent Reactor Kinetics," *Nucl. Sci. Eng.* **31,** 67 (1968).
14. W. E. Kastenberg, "A Stability Criterion for Space-Dependent Nuclear-Reactor Systems with Variable Temperature Feedback," *Nucl. Sci. Eng.* **37,** 19 (1969).

Chapter 7

CONTROL

The control of a nuclear reactor to optimize its performance is a problem that at present is solved on an intuitive or trial and error basis. However, the manner in which a reactor is controlled can significantly affect its performance,† and it is likely that increasing attention will be paid to this question in the future. The methods of modern control theory that have been developed extensively in other disciplines constitute a theoretical basis that may be specialized and extended to the needs of nuclear reactor control.

This chapter is concerned with the variational theory that is the basis for the better-known methods of modern control theory. Application of the calculus of variations and the use of dynamic programming to derive Pontryagin's maximum principle are discussed for both ordinary and partial differential equation formulations of space-time nuclear reactor dynamics. An application of variational synthesis methods in solving optimal control problems is outlined.

† An example already mentioned in Chapter 5, xenon spatial oscillations, serves as an illustration. Most reactor cores are designed to achieve a relatively uniform power distribution, and the maximum power level is set by a maximum allowable power density. If spatial flux tilts produce a 25% increase in power density at a limiting location, then the maximum power density must be reduced by 25%. Alternatively, the core must be designed to withstand 25% higher power densities than nominal, which is expensive. In either case, it may be more economical, if not imperative, to control to eliminate spatial flux tilting.

7.1 Variational Methods of Modern Control Theory

When the approximation methods of Chapter 1 are employed, the dynamics of a spatially dependent nuclear reactor model are described by a system of ordinary differential equations

$$\dot{y}_i(t) = f_i(y_1, \ldots, y_N, u_1, \ldots, u_R), \qquad i = 1, \ldots, N, \tag{7.1}$$

with the initial conditions

$$y_i(t = t_0) = y_{i0}, \qquad i = 1, \ldots, N. \tag{7.2}$$

The y_i are the state variables (e.g., nodal neutron flux, temperature) and the u_r are control variables (e.g., control rod cross section in a node). Equations (7.1) may be written more compactly by defining vector variables **y, u,** and **f**

$$\dot{\mathbf{y}}(t) = \mathbf{f}(\mathbf{y}(t), \mathbf{u}(t)). \tag{7.3}$$

Many problems in optimal control may be formulated as a quest for the control vector $\mathbf{u}^*$ that causes the solution of Equation (7.3), $\mathbf{y}^*$, to minimize a functional†

$$J[\mathbf{y}] = \int_{t_0}^{t_f} dt\, F(\mathbf{y}(t), \dot{\mathbf{y}}(t)). \tag{7.4}$$

This optimal control problem may be formulated within the framework of the classical calculus of variations by treating the control variables as equivalent to the state variables. The theory of the calculus of variations[1] is restricted to variables that are continuous in time, which limits the admissible set of control variables.

The system equations are treated as constraints or subsidiary conditions, and are included in the functional with Lagrange multiplier variables

$$J'[\mathbf{y}, \mathbf{u}, \boldsymbol{\lambda}] = \int_{t_0}^{t_f} dt \left[F(\mathbf{y}(t), \dot{\mathbf{y}}(t)) + \sum_{i=1}^{N} \lambda_i(t)\{\dot{y}_i(t) - f_i(\mathbf{y}(t), \mathbf{u}(t))\}\right]. \tag{7.5}$$

† Functionals of this form may arise when the objective of the control program is to correct a flux perturbation in such a manner as to minimize the deviation from the nominal flux distribution, at the same time minimizing the rate of change of local flux densities. Other typical control problems are those in which the objective is to attain a given final state in a minimum time; a functional with $F = 1$ and an additional term that provides a measure of the deviation from the specified final state is appropriate in this case. Such problems are referred to as terminal control problems and are discussed by Dreyfus,[2] Fel'dbaum,[3] and Pontryagin *et al.*[4]

Variations of the modified functional J' (with respect to each y_i and u_r) are required to vanish at the minimum

$$\delta J' = 0 = \int_{t_0}^{t_f} dt \left[\sum_{i=1}^{N} \left\{ \frac{\partial F}{\partial y_i} \delta y_i + \frac{\partial F}{\partial \dot{y}_i} \delta \dot{y}_i + \lambda_i \, \delta \dot{y}_i - \sum_{j=1}^{N} \lambda_j \frac{\partial f_j}{\partial y_i} \delta y_i - \sum_{r=1}^{R} \lambda_i \frac{\partial f_i}{\partial u_r} \delta u_r \right\} \right]. \tag{7.6}$$

Integrating the $\delta\dot{y}_i$ terms by parts,† and using the initial conditions to set $\delta y_i(t_0) = 0$, this expression becomes

$$\delta J' = 0 = \int_{t_0}^{t_f} dt \sum_{i=1}^{N} \left[\left\{ \frac{\partial F}{\partial y_i} - \frac{\partial}{\partial t}\frac{\partial F}{\partial \dot{y}_i} - \dot{\lambda}_i - \sum_{j=1}^{N} \lambda_i \frac{\partial f_j}{\partial y_i} \right\} \delta y_i - \sum_{r=1}^{R} \lambda_i \frac{\partial f_i}{\partial u_r} \delta u_r \right] + \sum_{i=1}^{N} \left(\lambda_i + \frac{\partial F}{\partial \dot{y}_i} \right) \delta y_i \Bigg|_{t=t_f} . \tag{7.7}$$

In order that Equation (7.7) be satisfied for arbitrary (but continuous) variations δy_i and δu_r, it is necessary that

$$\dot{\lambda}_i(t) = -\sum_{j=1}^{N} \lambda_j(t) \frac{\partial f_j}{\partial y_i}(\mathbf{y}, \mathbf{u}) + \left\{ \frac{\partial F}{\partial y_i}(\mathbf{y}, \mathbf{f}) - \frac{\partial}{\partial t}\frac{\partial F}{\partial \dot{y}_i}(\mathbf{y}, \mathbf{f}) \right\}, \qquad i = 1, \ldots, N, \tag{7.8}$$

$$\sum_{i=1}^{N} \lambda_i(t) \frac{\partial f_i}{\partial u_r}(\mathbf{y}, \mathbf{u}) = 0, \qquad r = 1, \ldots, R, \tag{7.9}$$

and that λ_i satisfy the final conditions

$$\lambda_i(t_f) + \frac{\partial F}{\partial \dot{y}_i}\bigg|_{t_f} = 0, \qquad i = 1, \ldots, N. \tag{7.10}$$

Equations (7.1), (7.8), and (7.9) must be solved simultaneously, subject to the initial conditions of Equations (7.2) and the final conditions of Equations (7.10), for the optimal controls $u_r{}^*(t)$ and the optimal solutions $y_i{}^*(t)$.

In many problems additional constraints are placed upon the allowable values that may be taken on by the state variables and control variables.

† Implying the operators δ and d/dt commute, and that δy_i are continuous in time.

Constraints of the form

$$\phi_m(\mathbf{y}(t), \mathbf{u}(t)) = 0, \qquad m = 1, \ldots, M < N,$$

or

$$\phi_m(\mathbf{y}(t), \dot{\mathbf{y}}(t), \mathbf{u}(t), \dot{\mathbf{u}}(t)) = 0, \qquad m = 1, \ldots, M < N,$$

may be added to the functional of Equation (7.4) with Lagrange multiplier variables and treated in the same fashion as before. Equations for additional Lagrange multiplier variables and the additional constraint equations are included with Equations (7.1), (7.8), and (7.9) in this case.

When integral constraints of the form

$$\int_{t_0}^{t_f} dt\, \phi_m(\mathbf{y}(t), \dot{\mathbf{y}}(t), \mathbf{u}(t), \dot{\mathbf{u}}(t)) = l_m, \qquad m = 1, \ldots, M < N,$$

are present, the functional of Equation (7.4) is modified with Lagrange multiplier constants ω_m,

$$J \to \hat{J} = \int_{t_0}^{t_f} dt \left[F + \sum_{m=1}^{M} \omega_m \phi_m \right] = \int_{t_0}^{t_f} dt\, \hat{F},$$

and the derivation proceeds as before with $F \to \hat{F}$. In addition to Equations (7.1), (7.8), and (7.9), the constraint equations and expressions for the ω_m are obtained.

Inequality constraints (e.g , maximum control rod shim rates) are encountered frequently. Although these can sometimes be reduced to equivalent equality constraints of one of the three types discussed, they generally constitute a class of problems that are difficult to treat within the framework of the calculus of variations. Another class of such problems is those for which the optimal control is discontinuous.

An alternate treatment of the variational problem, posed at the beginning of this section, that circumvents the requirement for continuous control variables is provided by dynamic programming.[5] Consider the problem of determining the control vector $\mathbf{u}^*(t)$ that causes the solution $\mathbf{y}^*(t)$ of Equation (7.3) to minimize the functional of Equation (7.4), subject to constraints on the allowable values of the control variable that may be represented by

$$\mathbf{u}(t) \in \Omega(t). \tag{7.11}$$

To develop the dynamic programming formalism, consider the functional of Equation (7.4) evaluated between a variable lower limit $(t, \mathbf{y}(t))$ and a fixed upper limit $(t_f, \mathbf{y}(t_f))$. Define the minimum value of this functional as S, a function of the lower, variable limit $(t, \mathbf{y}(t))$

$$S(t, \mathbf{y}(t)) \equiv \min_{\mathbf{u}\in\Omega} \int_t^{t_f} dt'\, F(\mathbf{y}(t'), \dot{\mathbf{y}}(\mathbf{y}(t'), \mathbf{u}(t'))). \tag{7.12}$$

In writing Equation (7.12), $\dot{\mathbf{y}}$ is written as an explicit function of $\mathbf{y}$ and $\mathbf{u}$ to indicate that Equation (7.3) must be satisfied in evaluating the integrand.

By definition of S, for $\Delta t > 0$

$$S(t, \mathbf{y}(t)) \leqq S(t + \Delta t, \mathbf{y}(t + \Delta t)) + \int_t^{t+\Delta t} dt'\, F(\mathbf{y}(t'), \dot{\mathbf{y}}(\mathbf{y}(t'), \mathbf{u}(t'))), \tag{7.13}$$

where $\mathbf{y}(t + \Delta t)$ and $\mathbf{y}(t)$ are related by Equation (7.3); i.e.,

$$\mathbf{y}(t + \Delta t) = \mathbf{y}(t) + \Delta t\, \mathbf{f}(\mathbf{y}(t), \mathbf{u}(t)) + O(\Delta t^2). \tag{7.14}$$

For the optimal choice of $\mathbf{u}(t') = \mathbf{u}^*$ in the interval $t \leqq t' \leqq t + \Delta t$, the equality obtains in Equation (7.13). Approximating the integral in Equation (7.13) by taking the integrand constant at its value at t, this equation becomes†

$$S(t, \mathbf{y}(t)) = \min_{\mathbf{u}(t)\in\Omega(t)} [S(t + \Delta t, \mathbf{y}(t + \Delta t)) + \Delta t\, F(\mathbf{y}(t), \mathbf{f}(\mathbf{y}(t), \mathbf{u}(t)))]. \tag{7.15}$$

Equation (7.15) can be solved by retrograde calculation, starting with the final condition

$$S(t_f, \mathbf{y}(t_f)) = \min_{\mathbf{u}\in\Omega} \int_{t_f}^{t_f} dt'\, F(\mathbf{y}, \dot{\mathbf{y}}) = 0. \tag{7.16}$$

In each step of the retrograde solution, the optimal manner to proceed from each possible state $\mathbf{y}(t)$ to time t_f is computed. Thus, when the initial time is reached, the optimal control at each discrete time and the corresponding sequence of states constituting the optimal trajectory are known. The dynamic programming method is discussed exhaustively by Bellman[5] and Dreyfus.[2]

This formalism is identical to the computing algorithm developed from physical arguments in Section 5.4 [i.e., Equation (5.54)].

† In Equation (7.15), the minimization is with respect to the values of the control vector at time t. These values are assumed constant over the interval t to $t + \Delta t$. On the other hand, the minimization in Equation (7.12) is with respect to the values taken on by the control vector at all times t', $t \leqq t' \leqq t_f$.

When a Taylor's series expansion of the first term on the right of Equation (7.15) is made,† this equation becomes

$$0 = \min_{\mathbf{u}(t)\in\Omega(t)} \left[\frac{\partial S}{\partial t}(t, \mathbf{y}(t)) + \sum_{i=1}^{N} \frac{\partial S}{\partial y_i}(t, \mathbf{y}(t)) f_i(\mathbf{y}(t), \mathbf{u}(t)) + F(\mathbf{y}(t), \mathbf{f}(\mathbf{y}(t), \mathbf{u}(t))) \right]. \tag{7.17}$$

Define the variables

$$\Psi_i(t) \equiv -\frac{\partial S}{\partial y_i}(t, \mathbf{y}(t)), \qquad i = 1, \ldots, N, \tag{7.18}$$

$$\Psi_{N+1}(t) \equiv -\frac{\partial S}{\partial t}(t, \mathbf{y}(t)). \tag{7.19}$$

With these definitions, Equation (7.17) becomes

$$0 = \min_{\mathbf{u}(t)\in\Omega(t)} \left[-\Psi_{N+1}(t) - \sum_{i=1}^{N} \Psi_i(t) f_i(\mathbf{y}(t), \mathbf{u}(t)) + F(\mathbf{y}(t), \mathbf{f}(\mathbf{y}(t), \mathbf{u}(t))) \right],$$

which may be written

$$0 = \max_{\mathbf{u}(t)\in\Omega(t)} \left[\Psi_{N+1}(t) + \sum_{i=1}^{N} \Psi_i(t) f_i(\mathbf{y}(t), \mathbf{u}(t)) - F(\mathbf{y}(t), \mathbf{f}(\mathbf{y}(t), \mathbf{u}(t))) \right]. \tag{7.20}$$

This is the maximum principle of Pontryagin.[4] When the vector $\mathbf{u}(t)$ takes on its optimal value, derivatives of the quantity within the square bracket with respect to t and y_i must vanish, which requires that

$$\frac{\partial \Psi_{N+1}}{\partial y_j} + \sum_{i=1}^{N} \frac{\partial \Psi_i}{\partial y_j} f_i = -\sum_{i=1}^{N} \Psi_i \frac{\partial f_i}{\partial y_j} + \left(\frac{\partial F}{\partial y_j} + \sum_{i=1}^{N} \frac{\partial F}{\partial \dot{y}_i} \frac{\partial f_i}{\partial y_j} \right), \qquad j = 1, \ldots, N,$$

$$\frac{\partial \Psi_{N+1}}{\partial t} + \sum_{i=1}^{N} \frac{\partial \Psi_i}{\partial t} f_i = -\sum_{i=1}^{N} \Psi_i \frac{\partial f_i}{\partial t}.$$

† This implies that partial derivatives of S with respect to t and the y_i exist. An alternate development of the subsequent material that avoids this assumption may be found on pp. 98–105 of Fel'dbaum.[3]

Using the identities

$$\frac{d\Psi_j}{dt} = -\sum_{i=1}^{N} \frac{\partial^2 S}{\partial y_i\, \partial y_j} f_i - \frac{\partial^2 S}{\partial t\, \partial y_j} = \sum_{i=1}^{N} \frac{\partial \Psi_i}{\partial y_j} f_i + \frac{\partial \Psi_{N+1}}{\partial y_j}, \quad j = 1, \ldots, N,$$

and

$$\frac{d\Psi_{N+1}}{dt} = -\sum_{i=1}^{N} \frac{\partial^2 S}{\partial y_i\, \partial t} f_i - \frac{\partial^2 S}{\partial t^2} = \sum_{i=1}^{N} \frac{\partial \Psi_i}{\partial t} f_i + \frac{\partial \Psi_{N+1}}{\partial t},$$

these equations become

$$\frac{d\Psi_j}{dt} = -\sum_{i=1}^{N} \Psi_i \frac{\partial f_i}{\partial y_j} + \left(\frac{\partial F}{\partial y_j} + \sum_{i=1}^{N} \frac{\partial F}{\partial \dot{y}_i} \frac{\partial f_i}{\partial y_j}\right), \quad j = 1, \ldots, N, \tag{7.21}$$

$$\frac{d\Psi_{N+1}}{dt} = -\sum_{i=1}^{N} \Psi_i \frac{\partial f_i}{\partial t}. \tag{7.22}$$

Appropriate final conditions for the Ψ_i and Ψ_{N+1} can be shown (Fel'dbaum,[3] p. 102) to be

$$\Psi_1(t_f) = \cdots = \Psi_N(t_f) = \Psi_{N+1}(t_f) = 0. \tag{7.23}$$

Thus, Equations (7.1), (7.20), (7.21), and (7.22) are solved simultaneously, subject to the initial and final conditions of Equations (7.2) and (7.23), respectively.

The computational procedure for solving either the calculus of variations or maximum principle equations is generally iterative. At $t = t_0$, the y_i are known from the initial conditions. When the maximum principle formulation is used, initial values of Ψ_i are guessed, and the initial value of the control variables are determined from Equation (7.20). Then the y_i and Ψ_i are calculated at $t_0 + \Delta t$ from Equations (7.1) and (7.21) and (7.22), and the control is found from Equation (7.20), etc. This procedure is repeated in small time increments until the final time t_f. Then $\Psi_i(t_f)$ and $\Psi_{N+1}(t_f)$ are compared with the final conditions

$$\Psi_1(t_f) = \cdots = \Psi_N(t_f) = \Psi_{N+1}(t_f) = 0,$$

and the initial values of Ψ_i and Ψ_{N+1} are changed and the whole process is repeated. This is continued until a set of initial values $\Psi_i(t_0)$ and $\Psi_{N+1}(t_0)$ are found that yield the correct final values. A similar procedure is followed to solve the equations of the calculus of variations formulation.

These methods have not been used widely for spatially dependent nuclear reactor kinetics problems. Use of the dynamic programming algorithm of Equation (7.15) in determining the optimal control of xenon spatial oscillations was discussed in Section 5.4. Application of these, and other, techniques to the optimal control of transients in idealized reactor models is discussed by Ash[6] and Weaver.[7]

7.2 Variational Synthesis of Optimal Solutions

The techniques of variational synthesis (see Chapter 3) can be applied to obtain approximate solutions of optimal control problems. In general, the optimality functional will not depend on all the state variables. Assuming that the optimality functional depends upon $y_1, y_2, \ldots, y_L$, $L < N$, define the vector variables $\mathbf{x} = (y_1, \ldots, y_L)$, $\mathbf{z} = (y_{L+1}, \ldots, y_N)$, $\mathbf{f}_x = (f_1, \ldots, f_L)$, and $\mathbf{f}_z = (f_{L+1}, \ldots, f_N)$. Write Equations (7.1) and (7.2) as

$$\dot{\mathbf{x}}(t) = \mathbf{f}_x(\mathbf{x}(t), \mathbf{z}(t), \mathbf{u}(t)), \qquad \mathbf{x}(0) = \mathbf{x}_0 \tag{7.24}$$

$$\dot{\mathbf{z}}(t) = \mathbf{f}_z(\mathbf{x}(t), \mathbf{z}(t), \mathbf{u}(t)), \qquad \mathbf{z}(0) = \mathbf{z}_0 \tag{7.25}$$

and write the optimality functional (the minimization of which is the objective of the control program)

$$J[\mathbf{x}, \mathbf{u}] = \int_0^{t_f} dt\, F(\mathbf{x}(t), \mathbf{u}(t)). \tag{7.26}$$

(The lower time limit t_0 has been replaced by 0 to simplify the subsequent notation.)

To solve the control problem of choosing $\mathbf{u}^*$ to minimize J subject to Equations (7.24) and (7.25), Equation (7.24) is first rearranged,

$$\dot{\mathbf{x}}(t) = \mathbf{A}\mathbf{x}(t) + \hat{\mathbf{u}}(t), \tag{7.27}$$

where $\mathbf{A}$ is an $L \times L$ time-independent matrix, and

$$\hat{\mathbf{u}}(t) = \mathbf{f}_x(\mathbf{x}(t), \mathbf{z}(t), \mathbf{u}(t)) - \mathbf{A}\mathbf{x}(t). \tag{7.28}$$

Equation (7.27) has the formal solution

$$\mathbf{x}(t) = \exp(-\,t\mathbf{A})\mathbf{x}_0 + \int_0^t dt'\, \exp(-(t - t')\mathbf{A})\hat{\mathbf{u}}(t'). \tag{7.29}$$

When $\mathbf{x}$, $\mathbf{x}_0$, and $\hat{\mathbf{u}}$ are expanded in eigenvectors† $\boldsymbol{\Psi}_m$ of $\mathbf{A}$ (with eigenvalues ω_n)

$$\mathbf{x}(t) = \sum_{m=1}^{M} \boldsymbol{\Psi}_m x_m(t),$$

$$\mathbf{x}_0 = \sum_{m=1}^{M} \boldsymbol{\Psi}_m b_m,$$

$$\hat{\mathbf{u}}(t) = \sum_{m=1}^{M} \boldsymbol{\Psi}_m a_m(t), \tag{7.30}$$

and the biorthogonality condition

$$\boldsymbol{\Psi}_n^{+T} \boldsymbol{\Psi}_m = \delta_{n,m} \tag{7.31}$$

is imposed ($\boldsymbol{\Psi}_n^+$ are eigenvectors of the Hermitian adjoint of $\mathbf{A}$ with eigenvalues ω_n), Equation (7.29) can be used to relate the x_n and a_n,

$$x_n(t) \equiv \boldsymbol{\Psi}_n^{+T} \mathbf{x}(t) = e^{-\omega_n t} b_n + \int_0^t dt' \, e^{-\omega_n(t-t')} a_n(t'), \tag{7.32}$$

$$n = 1, \ldots, M.$$

The $a_n(t)$ are next expanded in known functions of time, $\phi_k(t)$, with unknown constant coefficients,

$$a_n(t) = \sum_{k=1}^{K} d_{nk} \, \phi_k(t), \tag{7.33}$$

and Equations (7.32) become

$$x_n(t) = e^{-\omega_n t} b_n + \sum_{k=1}^{K} d_{nk} \int_0^t dt' \, e^{-\omega_n(t-t')} \phi_k(t'), \tag{7.34}$$

$$n = 1, \ldots, M.$$

When the expansions of Equations (7.30) are used in the functional of Equation (7.26), and Equations (7.34) are employed to incorporate the constraints imposed upon the optimal solution by Equations (7.24), a

† Expansion in other vector functions could be used, too. However, only when expansions in the eigenvectors of $\mathbf{A}$ are used do the equations corresponding to the different expansion vectors decouple and is it possible to represent $\exp(t\mathbf{A})$ so simply.

reduced functional depending on the d_{nk} is obtained. Equations for the d_{nk} are obtained by minimizing this reduced functional

$$\frac{\delta J}{\delta d_{nk}}[d_{11}, \dots, d_{1K}, \dots, d_{MK}] = 0, \tag{7.35}$$

$$n = 1, \dots, M, \quad k = 1, \dots, K.$$

Equations (7.35) are MK algebraic equations which can be solved for the d_{nk}.

A final condition $\mathbf{x}(t_f) = \mathbf{Q}$ can be incorporated by using Equations (7.34) to obtain a relationship among the d_{nk}

$$d_{m1} = \frac{\mathbf{\Psi}_m^{+T}\mathbf{Q} - e^{-\omega_m t_f} b_n - \sum_{k=2}^{K} d_{mk} \int_0^{t_f} dt\, e^{-(t_f - t)\omega_m} \phi_k(t)}{\int_0^{t_f} dt\, e^{-(t_f - t)\omega_m} \phi_1(t)}, \tag{7.36}$$

$$m = 1, \dots, M.$$

This relationship is used to eliminate d_{m1} from the reduced optimality functional. In this case, those of Equations (7.35) corresponding to $k = 1$ are replaced by Equations (7.36).

Thus, the MK algebraic equations are solved for the d_{mk}, then the optimal $\hat{\mathbf{u}}^*(t)$ and $\mathbf{x}^*(t)$ are constructed from Equations (7.30), (7.32), and (7.33). Knowing $\hat{\mathbf{u}}^*(t)$ and $\mathbf{x}^*(t)$, Equations (7.25) and (7.28) can be solved readily for the optimal control $\mathbf{u}^*(t)$ and $\mathbf{z}(t)$.

In contrast to the methods of the previous section, which require the solution of coupled differential equations with final and initial conditions, the use of variational synthesis and an integral equation formulation of the reactor dynamics reduces the optimization problem to the solution of a set of algebraic equations. Another salient feature of the method of this section is the formulation of the problem such that the optimal $\hat{\mathbf{u}}^*(t)$ is sought. This vector includes not only the control vector $\mathbf{u}$, but the other state variables $\mathbf{z}$ which are not needed in evaluating the optimality functional. By requiring the control vector to have the form given by Equations (7.30) and (7.32), further artificial constraints are imposed on the optimization problem, and $\mathbf{u}^*$ obtained in this manner may not be truly optimal.

7.3 Variational Methods for Spatially Dependent Control Problems

The basic description of the transient neutron flux and temperature distributions within a nuclear reactor is in terms of partial differential equations. It is not clear that the optimal control computed by first reducing these equations to ordinary differential equations by the methods of Chapter 1 and then using the methods of Section 7.1 is the same as would be obtained if the optimal control were determined directly from the partial differential equation description of the reactor dynamics. The variational formalism of Section 7.1 is extended to the partial differential equation description of the reactor dynamics in this section. The general problem of controlling systems described by partial differential equations is discussed by Wang,[8] and extensions of this theory to reactor control problems is discussed by Wiberg,[9] Hsu,[10] and Lewins and Babb.[11]

The state of the system is specified in terms of state functions $y_i(r, t)$, rather than discrete state variables as in Section 7.1. The function space Γ_i consisting of all possible functions of position that the state function y_i can take on is a component function space, and the product space $\Gamma = \Gamma_i \otimes \Gamma_2 \cdots \otimes \Gamma_N$ of all such component function spaces is the state function space on which the vector state function $\mathbf{y} = (y_1, \ldots, y_N)$ is defined. Similarly, the vector control function $\mathbf{u} = (u_1, \ldots, u_R)$ is defined on the product space of the component function spaces defined by all possible functions of position that the control functions u_r can take on. The "distance" between two states $\mathbf{y}_a$ and $\mathbf{y}_b$ is defined as the metric on Γ.

Equations for the reactor dynamics can be written in the form

$$\begin{aligned} \dot{y}_i(r, t) &= L_i(r) y_i(r, t) + f_i(\mathbf{y}, \mathbf{u}), \\ y_i(r, t_0) &= y_{i0}(r), \\ y_i(R, t) &= 0, \qquad i = 1, \ldots, N, \end{aligned} \tag{7.37}$$

where y_i denotes a state function, L_i contains a spatial differential operator acting on y_i, and f_i is a spatially dependent function of $\mathbf{y}$ and $\mathbf{u}$. The outer boundary of the reactor is denoted by R. These equations may be written in matrix format

$$\begin{aligned} \dot{\mathbf{y}}(r, t) &= \mathbf{L}(r)\mathbf{y}(r, t) + \mathbf{f}(\mathbf{y}, \mathbf{u}), \\ \mathbf{y}(r, 0) &= \mathbf{y}_0(r), \quad \mathbf{y}(R, t) = 0. \end{aligned} \tag{7.38}$$

Many optimal control problems may be formulated as the quest for the control vector function **u** for which the solution of Equation (7.38) minimizes a functional

$$J[\mathbf{y}, \mathbf{u}] = \int_{t_0}^{t_f} dt \int_V dr\, F(\mathbf{y}(r, t), \dot{\mathbf{y}}(r, t)). \tag{7.39}$$

The standard calculus of variations formulation of this problem begins by adding Equation (7.38) to the integrand of Equation (7.39) with a Lagrange multiplier vector function $\boldsymbol{\lambda}(r, t) = (\lambda_i, \ldots, \lambda_N)$.

$$J'[\mathbf{y}, \mathbf{u}, \boldsymbol{\lambda}] = \int_{t_0}^{t_f} dt \int_V dr\, [F(\mathbf{y}, \dot{\mathbf{y}}) + \boldsymbol{\lambda}^T(\dot{\mathbf{y}} - \mathbf{L}\mathbf{y} - \mathbf{f}(\mathbf{y}, \mathbf{u}))]. \tag{7.40}$$

The control functions u_r are treated in the same fashion as the state functions y_i. Next, the variation of J' is required to vanish.†

$$\delta J' = \int_{t_0}^{t_f} dt \int_V dr \sum_{i=1}^{N} \left[\frac{\partial F}{\partial y_i} \delta y_i + \frac{\partial F}{\partial \dot{y}_i} \delta \dot{y}_i \right.$$
$$\left. + \lambda_i\, \delta \dot{y}_i - \lambda_i L_i\, \delta y_i - \sum_{j=1}^{N} \lambda_i \frac{\partial f_i}{\partial y_j} \delta y_j - \sum_{r=1}^{R} \lambda_i \frac{\partial f_i}{\partial u_r} \delta u_r \right] = 0.$$

Integration by parts of the terms involving $\delta \dot{y}_i$ and $L_i\, \delta y_i$,‡ and use of the initial conditions $\delta y_i(r, t_0) = 0$ leads to

$$\delta J' = \int_{t_0}^{t_f} dt \int_V dt \sum_{i=1}^{N} \left[\delta y_i \left(\frac{\partial F}{\partial y_i} - \frac{\partial}{\partial t} \frac{\partial F}{\partial \dot{y}_i} - \dot{\lambda}_i - L_i^{+} \lambda_i - \sum_{j=1}^{N} \lambda_j \frac{\partial f_j}{\partial y_i} \right) \right.$$
$$\left. - \sum_{r=1}^{R} \lambda_i \frac{\partial f_i}{\partial u_r} \delta u_r \right] + \int_V dr \sum_{i=1}^{N} \frac{\partial F}{\partial \dot{y}_i} \delta y_i \bigg|_{t=t_f}$$
$$+ \int_V dr \sum_{i=1}^{N} \lambda_i\, \delta y_i \bigg|_{t=t_f} + \int_{t_0}^{t_f} dt \sum_{i=1}^{N} P_i(\lambda_i) = 0. \tag{7.41}$$

In arriving at Equation (7.41), the adjoint operator L_i^{+} and the bilinear concommitant P_i are defined by the relation

$$\int_V dr\, \lambda_i L_i\, \delta y_i = \int_V dr\, \delta y_i\, L_i^{+} \lambda_i + P_i(\lambda_i). \tag{7.42}$$

† Differentiation with respect to a function (e.g., $\partial F/\partial y_i$) is defined as a natural extension of the concept of a derivative (see Gelfand and Fomin,[1] pp. 11–14), so that use of the same notation should not cause any confusion. Continuously differentiable functions are assumed.

‡ Commutability of the variational operator δ and the operators $\partial/\partial t$ and L_i imply an assumption of continuous variations δy_i, as does the existence of the integrals involving these terms.

$\delta J'$ must vanish for arbitrary variations of y_i and u_r, which requires that the Lagrange multiplier functions satisfy the partial differential equations

$$\dot{\lambda}_i(r,t) = -L_i^{+}(r)\lambda_i(r,t) - \sum_{j=1}^{N} \lambda_j(r,t)\frac{\partial f_j}{\partial y_i}(\mathbf{y},\mathbf{u}) + \left(\frac{\partial F}{\partial y_i}(\mathbf{y},\dot{\mathbf{y}}) - \frac{\partial}{\partial t}\frac{\partial F}{\partial \dot{y}_i}(\mathbf{y},\dot{\mathbf{y}})\right), \tag{7.43}$$

$$i = 1, \ldots, N,$$

the final conditions

$$\left(\lambda_i + \frac{\partial F}{\partial \dot{y}_i}\right)_{t=t_f} = 0, \qquad i = 1, \ldots, N \tag{7.44}$$

and the boundary conditions

$$P_i(\lambda_i(R,t)) = 0, \qquad i = 1, \ldots, N. \tag{7.45}$$

In addition,

$$\sum_{i=1}^{N} \lambda_i(r,t)\frac{\partial f_i}{\partial u_r}(\mathbf{y},\mathbf{u}) = 0, \qquad r = 1, \ldots, R \tag{7.46}$$

must be satisfied.

In this formulation, the u_r, as well as the y_i, are treated as continuous functions. This imposes artificial restrictions on the u_r. In some problems the optimal control is discontinuous.

Additional constraints can be incorporated into the development in a manner similar to that discussed in Section 7.1 for the treatment of constraints. This formalism has been applied to several idealized reactor control problems.[9]

Proceeding as in Section 7.1, the dynamic programming formalism is developed by considering the minimum value of the functional of Equation (7.39) evaluated between a fixed upper limit and a variable lower limit as a function of the lower limit:

$$S(t,\mathbf{y}(r,t)) = \min_{\mathbf{u}\in\Omega} \int_t^{t_f} dt' \int_V dr\, F(\mathbf{y}(r,t'), \dot{\mathbf{y}}(\mathbf{y}(r,t'), \mathbf{u}(r,t'))). \tag{7.47}$$

In writing Equation (7.47), the dependence of the integrand upon

Equation (7.38) is shown implicitly, and any constraints on the control vector function are implied by $\mathbf{u} \in \boldsymbol{\Omega}$. By definition,

$$S(t, \mathbf{y}(r, t)) \leqq S(t + \Delta t, \mathbf{y}(r, t + \Delta t)) + \int_t^{t+\Delta t} dt' \int_V dr\, F(\mathbf{y}(r, t'), \dot{\mathbf{y}}(\mathbf{y}(r, t), \mathbf{u}(r, t))).$$

For the optimal control, the equality obtains. Approximating the integral over time, this becomes†

$$S(t, \mathbf{y}(r, t)) = \min_{\mathbf{u}(t) \in \Omega(t)} \left[S(t + \Delta t, \mathbf{y}(r, t + \Delta t)) + \Delta t \int_V dr\, F(\mathbf{y}(r, t), \dot{\mathbf{y}}(\mathbf{y}(r, t), \mathbf{u}(r, t))) \right]. \quad (7.48)$$

Equation (7.48) is the dynamic programming algorithm for the partial differential equation description of the reactor dynamics. It is solved retrogressively, with the final condition

$$S(t_f, \mathbf{y}(r, t)) = 0, \quad (7.49)$$

which is apparent from the defining Equation (7.47).

Using a Taylor's series expansion

$$S(t + \Delta t, \mathbf{y}(r, t + \Delta t)) = S(t, \mathbf{y}(r, t)) + \Delta t \frac{\partial S}{\partial t}(t, \mathbf{y}(r, t)) + \Delta t \int_V dr \sum_{i=1}^{N} \frac{\partial S}{\partial y_i}(t, \mathbf{y}(r, t)) \times \{L_i(r) y_i(r, t) + f_i(\mathbf{y}(r, t), \mathbf{u}(r, t))\},$$

Equation (7.48) becomes

$$0 = \min_{\mathbf{u}(t) \in \Omega(t)} \left[\frac{\partial S}{\partial t}(t, \mathbf{y}(r, t)) + \sum_{i=1}^{N} \int_V dr \frac{\partial S}{\partial y_i}(t, \mathbf{y}(r, t)) \{L_i(r) y_i(r, t) + f_i(\mathbf{y}(r, t), \mathbf{u}(r, t))\} + \int_V dr\, F(\mathbf{y}(r, t), \dot{\mathbf{y}}(\mathbf{y}(r, t), \mathbf{u}(r, t))) \right]. \quad (7.50)$$

† The minimization in Equation (7.47) is with respect to the control vector function over the time interval $t \leqq t' \leqq t_f$, whereas the minimization in Equation (7.48) is with respect to the control vector function evaluated at time t.

Define the functions

$$\Psi_i(r, t) = -\frac{\partial S}{\partial y_i}(t, \mathbf{y}(r, t)), \qquad i = 1, \ldots, N, \tag{7.51}$$

$$\Psi_{N+1}(r, t) = -\frac{\partial S}{\partial t}(t, \mathbf{y}(r, t)). \tag{7.52}$$

Then Equation (7.50) becomes

$$0 = \max_{\mathbf{u}(t) \in \Omega(t)} \left[\Psi_{N+1}(r, t) + \sum_{i=1}^{N} \int_V dr\, \Psi_i(r, t)\{L_i(r) y_i(r, t) + f_i(\mathbf{y}(r, t), \mathbf{u}(r, t))\} - \int_V dr\, F(\mathbf{y}(r, t), \dot{\mathbf{y}}(\mathbf{y}(r, t), \mathbf{u}(r, t))) \right] \tag{7.53}$$

This is the extension of Pontryagin's maximum principle to the partial differential equation description of the reactor dynamics.

When the optimal $\mathbf{u}^*(t)$ is chosen, variational derivatives (Gelfand and Fomin,[1] pp. 27–29) of the quantity within the square brackets must vanish. This leads to the boundary condition

$$P_i(\Psi_i(R, t)) = 0 \tag{7.54}$$

where P_i is the bilinear concommitant defined in Equation (7.42), and to the equations

$$\frac{d\Psi_j}{dt} = -L_j^{+}\Psi_j - \sum_{i=1}^{N} \Psi_i \frac{\partial f_i}{\partial y_j} + \left(\frac{\partial F}{\partial y_j} + \sum_{i=1}^{N} \frac{\partial F}{\partial \dot{y}_i} \frac{\partial}{\partial y_i} \{L_i y_i + f_i\} \right), \tag{7.55}$$

$$j = 1, \ldots, N,$$

$$\frac{d\Psi_{N+1}}{dt} = -\sum_{i=1}^{N} \Psi_i \frac{\partial}{\partial t}\{L_i y_i + f_i\}. \tag{7.56}$$

Identities similar to those just before Equation (7.21) have been used in arriving at these equations. Appropriate final conditions for the Ψ_i and Ψ_{N+1} are (Hsu,[10] p. 146)

$$\Psi_1(t_f) = \cdots = \Psi_N(t_f) = \Psi_{N+1}(t_f) = 0. \tag{7.57}$$

The optimal control functions must be found by solving Equations (7.38), (7.53), and (7.54)–(7.56). As in Section 7.1, the initial conditions associated with the y_i and the final conditions associated with the Ψ_i and Ψ_{N+1} produce a system of equations that must, in general, be solved iteratively. This formulation allows discontinuous control functions and can incorporate constraints on the control functions readily, which are its principle advantages with respect to the calculus of variations formulation presented in the first part of this section. These methods have been applied to several idealized reactor control problems.[10]

7.4 Control of Xenon Spatial Oscillations

In order to illustrate the use of the methods described in Sections 7.1 and 7.3, and to present an alternate to the approach given in Section 5.4, the problem of controlling xenon-induced spatial power oscillations is formulated in terms of the calculus of variations.

When the spatial dependence is represented by the nodal approximation, a general optimality functional of the form of Equation (7.4) may be written (for a M-node model)

$$J[\phi_1, \ldots, \phi_M, u_1, \ldots, u_M] = \sum_{m=1}^{M} \int_{t_0}^{t_f} dt\, [(\phi_m(t) - N_m(t))^2 + K u_m{}^2(t)], \tag{7.58}$$

where ϕ_m and N_m represent the actual and the desired, respectively, time-dependent fluxes in node m, u_m is the control in node m, and K is a constant which can be varied to influence the relative importance of the two types of terms in the optimality functional. The purpose of the control program is to find the $u_m(t)$ which minimizes the optimality functional, subject to the constraints that the reactor remain critical,

$$\begin{aligned} 0 &= \sum_{m' \neq m}^{M} l_{mm'}(\phi_{m'}(t) - \phi_m(t)) + (\nu\Sigma_{fm} - \Sigma_{am} - \sigma_x X_m(t) - u_m(t))\phi_m(t) \\ &= f_{1m}, \end{aligned} \tag{7.59}$$

and the iodine and xenon dynamics equations are satisfied,

$$\begin{aligned} \dot{I}_m(t) &= \gamma_i \Sigma_{fm} \phi_m(t) - \lambda_i I_m(t) = f_{2m}, \\ I_m(0) &= I_{m0} \end{aligned} \tag{7.60}$$

$$\begin{aligned} \dot{X}_m(t) &= \gamma_x \Sigma_{fm} \phi_m(t) + \lambda_m I_m(t) - (\lambda_x + \sigma_x \phi_m(t)) X_m(t) = f_{3m}, \\ X_m(0) &= X_{m0}, \end{aligned} \tag{7.61}$$

$$m = 1, \ldots, M.$$

The m subscript denotes node m, $l_{mm'}$ is the internodal coupling coefficient discussed in Section 1.4, and the other quantities were defined in Chapter 5.

Equations (7.8) are

$$0 = 2(\phi_m(t) - N_m(t)) - (\nu\Sigma_{fm} - \Sigma_{am} - \sigma_x X_m(t) - u_m(t))\omega_{1m}(t) + \sum_{m' \neq m}^{M} l_{mm'}(\omega_{1m}(t) - \omega_{1m'}(t)) - \gamma_i \Sigma_{fm}\omega_{2m}(t) - (\gamma_x \Sigma_{fm} - \sigma_x X_m(t))\omega_{3m}(t), \tag{7.62}$$

$$\dot{\omega}_{2m}(t) = \lambda_i(\omega_{2m}(t) - \omega_{3m}(t)), \tag{7.63}$$

$$\dot{\omega}_{3m}(t) = \sigma_x \phi_m(t)\omega_{1m}(t) + (\lambda_x + \sigma_x \phi_m(t))\omega_{3m}(t), \tag{7.64}$$

$$m = 1, \ldots, M.$$

(The symbol ω has been used to denote the Lagrange multipliers, since λ is conventionally used to represent the decay constants.) The final conditions corresponding to Equations (7.10) are

$$\omega_{2m}(t_f) = 0, \quad \omega_{3m}(t_f) = 0, \qquad m = 1, \ldots, M. \tag{7.65}$$

Equations (7.9) are modified somewhat in this case because the optimality functional depends upon the control. The more general relation is

$$\frac{\partial F}{\partial u_r} + \sum_{i=1}^{N} \lambda_i(t) \frac{\partial f_i}{\partial u_r}(\mathbf{y}, \mathbf{u}) = 0, \qquad r = 1, \ldots, R, \tag{7.9'}$$

which becomes

$$2Ku_m(t) + \omega_{1m}(t)\phi_m(t) = 0, \qquad m = 1, \ldots, M. \tag{7.66}$$

Equations (7.66) can be used to eliminate the u_m from Equations (7.59) and (7.62). The modified equations, plus Equations (7.60), (7.61), (7.63), and (7.64) constitute a set of $6M$ equations which, together with the initial and final conditions specified above, can be solved for the optimal flux, iodine, xenon, and Lagrange multiplier trajectories. The optimal control can then be determined from Equations (7.66).

If no approximation is made for the spatial dependence, an equivalent optimality functional is

$$J[\phi, u] = \int_V dr \int_{t_0}^{t_f} dt\, [(\phi(r, t) - N(r, t))^2 + Ku^2(r, t)], \tag{7.67}$$

and the constraints are

$$0 = \nabla \cdot D(r)\, \nabla\phi(r,t) + (v\Sigma_f(r) - \Sigma_a(r) - \sigma_x X(r,t) - u(r,t))\phi(r,t) = f_1, \tag{7.68}$$

$$\dot{I}(r,t) = \gamma_i \Sigma_f(r)\phi(r,t) - \lambda_i I(r,t) = f_2, \quad I(r,0) = I_0(r) \tag{7.69}$$

$$\dot{X}(r,t) = \gamma_x \Sigma_f(r)\phi(r,t) + \lambda_i I(r,t) - (\lambda_x + \sigma_x\phi(r,t))X(r,t) = f_3, \quad X(r,0) = X_0(r). \tag{7.70}$$

Equations (7.43) become

$$\begin{aligned} 0 = {} & 2(\phi(r,t) - N(r,t)) - \nabla \cdot D(r)\, \nabla\omega_1(r,t) - (v\Sigma_f(r) - \Sigma_a(r) \\ & - \sigma_x X(r,t) - u(r,t))\omega_1(r,t) - \gamma_i \Sigma_f(r)\omega_2(r,t) \\ & - (\gamma_x \Sigma_f(r) - \sigma_x X(r,t))\omega_3(r,t), \end{aligned} \tag{7.71}$$

$$\dot{\omega}_2(r,t) = \lambda_i(\omega_2(r,t) - \omega_3(r,t)), \tag{7.72}$$

$$\dot{\omega}_3(r,t) = \sigma_x\phi(r,t)\omega_1(r,t) + (\lambda_x + \sigma_x\phi(r,t))\omega_3(r,t). \tag{7.73}$$

The final conditions of Equations (7.44) are

$$\omega_2(r, t_f) = \omega_3(r, t_f) = 0, \tag{7.74}$$

and the boundary condition of Equation (7.45) is

$$\omega_1(R, t) = 0. \tag{7.75}$$

Because the optimality functional contains the control functions, Equations (7.46) must be modified to

$$\frac{\partial F}{\partial u_r} + \sum_{i=1}^{N} \lambda_i(r, t) \frac{\partial f_i}{\partial u_r} (\mathbf{y}, \mathbf{u}) = 0, \tag{7.46$'$}$$

which becomes

$$2Ku(r,t) + \omega_1(r,t)\phi(r,t) = 0. \tag{7.76}$$

It is interesting to note that when the nodal approximation is made to Equations (7.68)–(7.76), they reduce identically to Equations (7.59)–(7.66). Thus, at least for this case, it does not matter whether the nodal

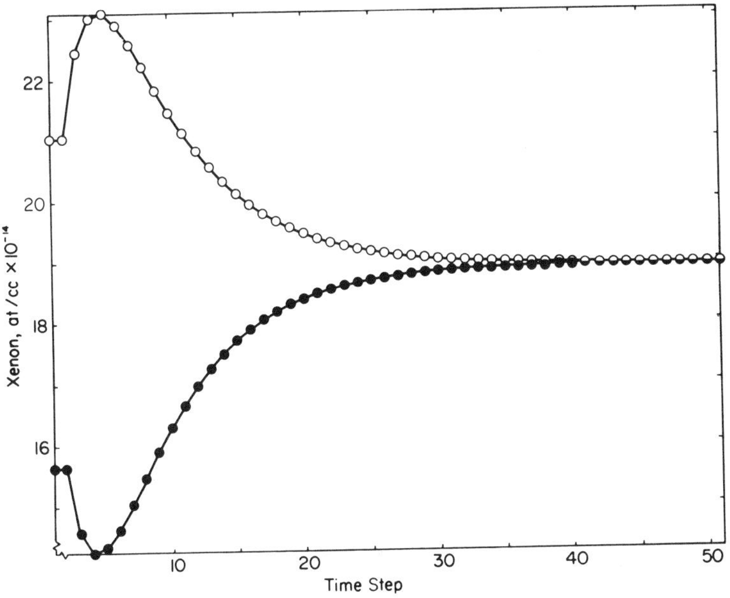

FIGURE 7.1. Optimal xenon solution, 2-node model.

approximation is made and then the equations for nodal Lagrange multipliers are derived according to Section 7.1, or the equations for the multipliers are derived according to the formalism of Section 7.3, and then the nodal approximation is made.

An iterative solution scheme for Equations (7.59)–(7.66) has been devised and applied to the simple case of a two-node reactor model. Figures 7.1–7.3 depict the optimal solution for a problem in which steady-state xenon and iodine concentrations were built up for an initial 2/1 flux tilt.† The objective of the control scheme was to eliminate the flux tilt and maintain a flux level of 1.5×10^{13} n/sec cm^2 in both nodes, which was accomplished by varying the control cross section to compensate the macroscopic xenon cross section in each node. (When $K = 0$ in the optimality functional, $u_m(t) = -\sigma_x X_m(t)$ is the obvious solution.) When K was increased so that the Ku^2 term was significant with respect to the $(\phi - N)^2$ term in the optimality functional, the amount of control was reduced somewhat, with the result that the desired flux level of

† The time step size was 2 hr for all results shown in this section.

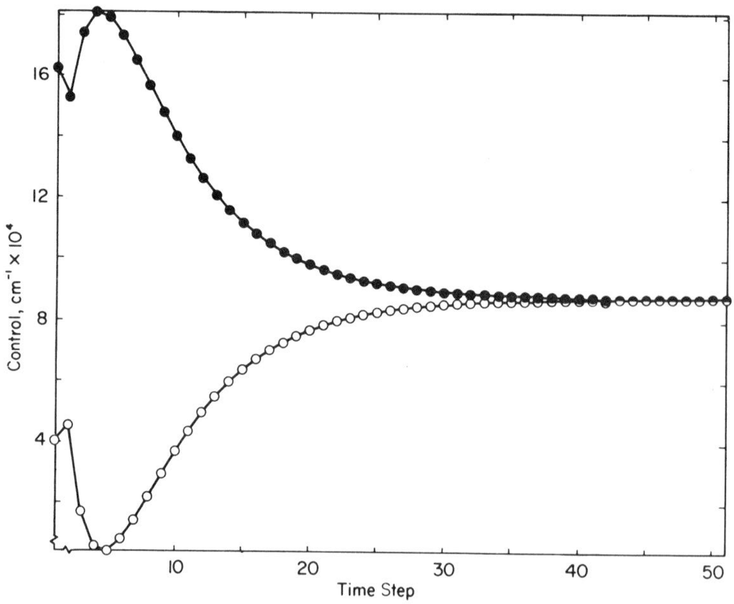

FIGURE 7.2. Optimal control, 2-node model.

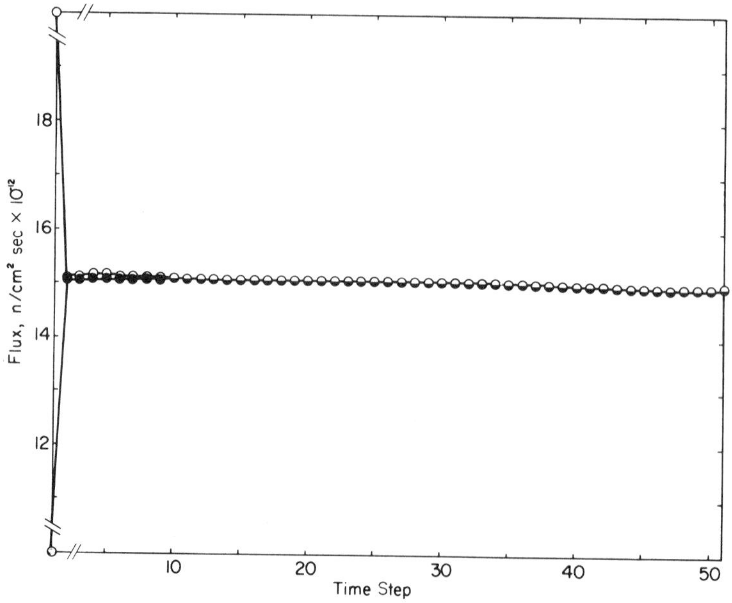

FIGURE 7.3. Optimal flux solution, 2-node model.

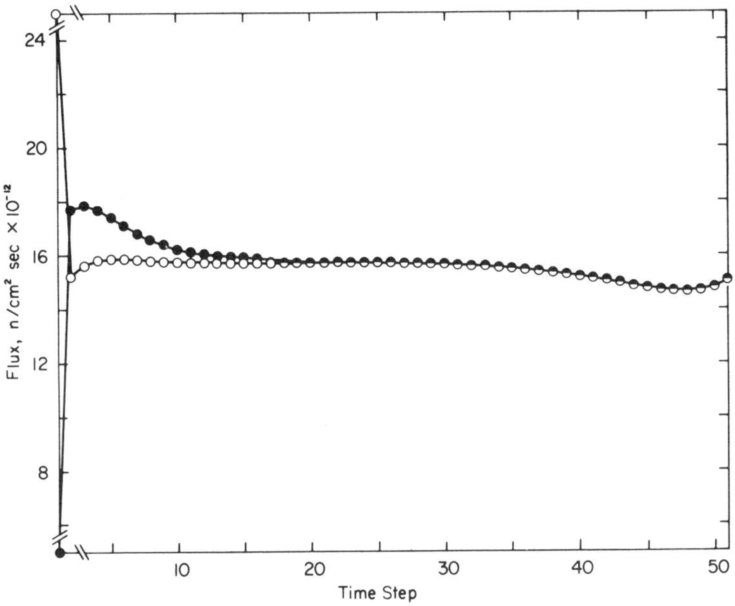

FIGURE 7.4. Optimal flux solution, 2-node model, increased K.

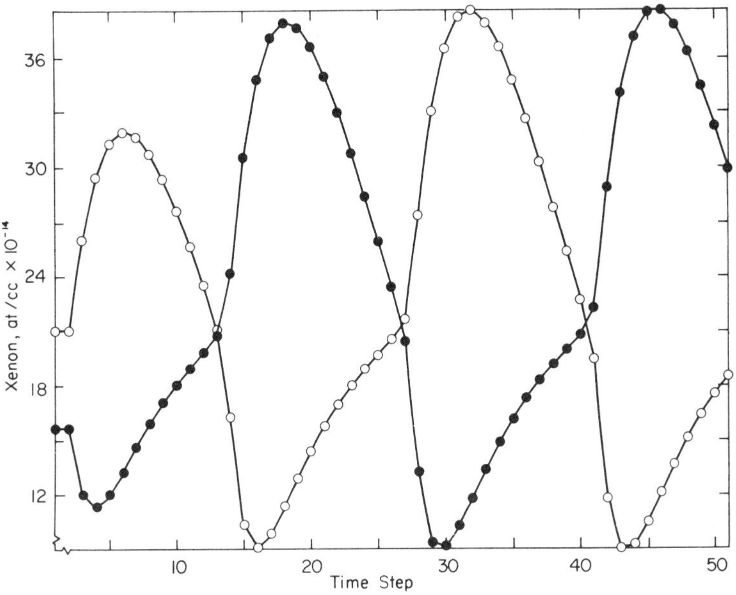

FIGURE 7.5. Xenon solution for uniform control, 2-node model.

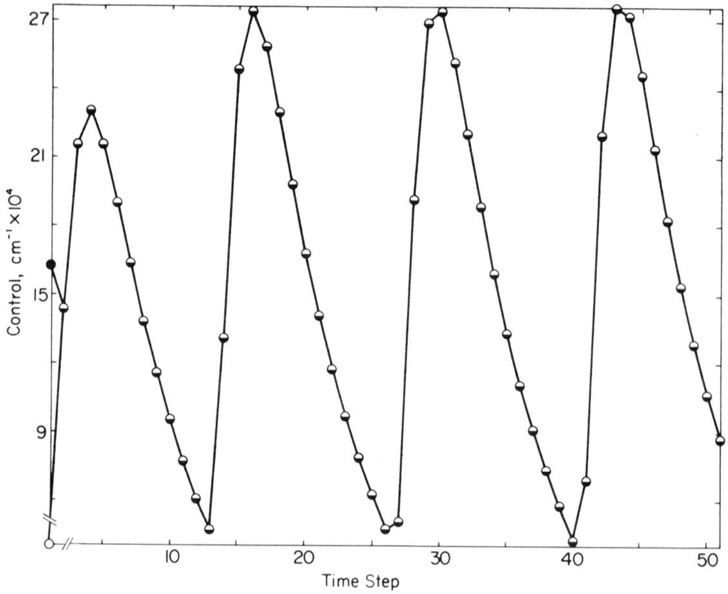

FIGURE 7.6. Uniform control required for criticality, 2-node model.

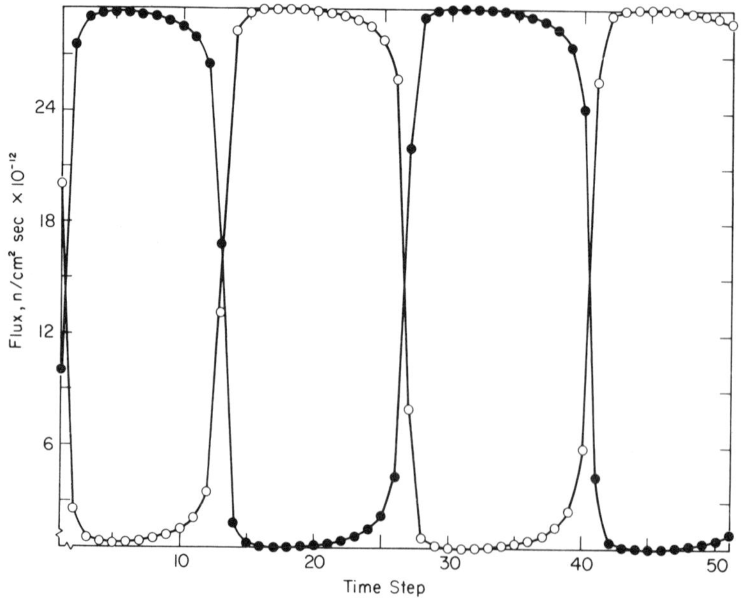

FIGURE 7.7. Limit cycle flux solution for uniform control, 2-node model.

1.5×10^{13} n/sec cm^2 in each node was not quite accomplished, as shown in Figure 7.4.

When the control was constrained to be the same in both nodes, simulating the motion of a uniform bank of control rods, the solution of the criticality and iodine–xenon dynamics equations was highly unstable, as shown in Figures 7.5–7.7. The flux is highly tilted to the node with the smallest xenon concentration, which eventually builds up the xenon concentration in that node until it is greater than the decaying xenon concentration in the other node, at which time the flux shifts dramatically to the other node. Since the reactivity effect of xenon and control is predominately due to the values of these variables in the high flux node, the control varies in such a manner as to compensate the reactivity effect of the xenon in the high flux node. As the flux shifts from the node in which the xenon concentration is growing to the node in which it is diminishing, the control passes through a minimum, and as the xenon concentration in the high flux node passes through a minimum, the control passes through a maximum. In more tightly coupled, less "tilty", models in which the xenon and control in the low flux node had a significant influence on the core reactivity, the variation in the control was less pronounced.

REFERENCES

1. I. M. Gelfand and S. V. Fomin, *Calculus of Variations.* Prentice-Hall, Englewood Cliffs, New Jersey, 1963.
2. S. E. Dreyfus, *Dynamic Programming and the Calculus of Variations.* Academic Press, New York, 1965.
3. A. A. Fel'dbaum, *Optimal Control Systems.* Academic Press, New York, 1965.
4. L. S. Pontryagin, V. G. Boltyanskii, R. V. Gamknelidze, and E. F. Mishchenko. *The Mathematical Theory of Optimum Processes.* Wiley (Interscience), New York, 1962.
5. R. Bellman, *Dynamic Programming.* Princeton University Press, Princeton, New Jersey, 1957.
6. M. Ash, *Optimal Shutdown Control of Nuclear Reactors.* Academic Press, New York, 1966.
7. L. E. Weaver *Reactor Dynamics and Control.* American Elsevier, New York, 1968.
8. P. K. C. Wang, "Control of Distributed Parameter Systems," *Advan. Control Systems,* **1,** 75 (1964).

9. D. M. Wiberg, "Optimal Control of Nuclear Reactor Systems," *Advan. Control Systems*, **5**, 301 (1967).
10. C. Hsu, "Control and Stability Analysis of Spatially Dependent Nuclear Reactor Systems," ANL-7322, Argonne National Laboratory (1967).
11. J. Lewins and A. L. Babb, "Optimum Nuclear Reactor Control Theory," *Advan. Nucl. Sci. Tech.* **4**, 252 (1968).

DERIVATION OF MULTIGROUP DIFFUSION EQUATIONS

The number of neutrons within an incremental volume element dr about position r with energy within the increment dE about energy E and with direction within the incremental solid angle $d\mathbf{\Omega}$ about the direction $\mathbf{\Omega}$ at time t is defined to be

$$N(r, E, \mathbf{\Omega}, t)\, dr\, dE\, d\mathbf{\Omega},$$

where $N(r, E, \mathbf{\Omega}, t)$ is the neutron distribution function. This distribution function contains all the information conventionally required about the neutronic state of the reactor.†

The interaction between neutrons and atomic nuclei is treated on a macroscopic basis by ignoring the details of the interaction process within the nucleus. Instead, cross sections are defined that specify the probability of a given reaction (e.g., neutron capture, elastic scattering) taking place, and the actual interaction is treated mathematically as occurring instantaneously.‡

A balance equation may be written upon an incremental volume by equating the time rate of change of the neutron density within the elemental

† $N(r, E, \mathbf{\Omega}, t)$ defines the distribution in space, energy, and direction of the mean value of the stochastic neutron distribution. Except for Chapter 4, this book is concerned solely with this mean value.

‡ This means that the resulting equations are not valid on the time scale of the neutron–nucleus interaction. Since the interaction time is many orders of magnitude smaller than the time scale of other phenomena of interest in reactor physics, this presents no problems.

volume $dr\, dE\, d\mathbf{\Omega}$ of phase space with the rate at which neutrons are being introduced into this volume element minus the rate at which they are being removed. Since reaction rates depend upon the product of the neutron density, the neutron speed v, and the cross section, it is convenient to define

$$F(r, E, \mathbf{\Omega}, t) \equiv vN(r, E, \mathbf{\Omega}, t).$$

The neutron balance equation may be written

$$\begin{aligned}
\frac{1}{v}\frac{\partial F}{\partial t}(r, E, \mathbf{\Omega}, t) &= -\mathbf{\Omega}\cdot\nabla F(r, E, \mathbf{\Omega}, t) - \Sigma_T(r, E, t)F(r, E, \mathbf{\Omega}, t) \\
&+ Q(r, E, \mathbf{\Omega}, t) + (1-\beta)\frac{\chi_P(E)}{4\pi} \\
&\times \int_0^\infty dE'\, \nu\Sigma_f(r, E', t) \int_{\Omega'} F(r, E', \mathbf{\Omega}', t)\, d\mathbf{\Omega}' \\
&+ \int_0^\infty dE' \int_{\Omega'} d\mathbf{\Omega}'\, \Sigma_s(r; E', \mathbf{\Omega}' \to E, \mathbf{\Omega}; t)F(r, E', \mathbf{\Omega}', t) \\
&+ \sum_{m=1}^{M} \lambda_m \frac{\chi_m(E)}{4\pi} C_m(r, t).
\end{aligned} \tag{A.1}$$

The first term describes the net diffusion of neutrons out of $dr\, dE\, d\mathbf{\Omega}$; $\mathbf{\Omega}$ is a unit vector in the direction $\mathbf{\Omega}$ and ∇ is the spatial gradient operator. The second term describes the rate at which neutrons are removed from $dr\, dE\, d\mathbf{\Omega}$ by absorption and scattering. An external source rate is denoted by Q. The rate at which prompt fission neutrons are introduced into $dr\, dE\, d\mathbf{\Omega}$ is given by the fourth term; the distribution of fission neutrons is assumed isotropic in direction. The fifth term defines the rate at which neutrons are scattered into $dr\, dE\, d\mathbf{\Omega}$. Delayed neutrons produced by the decay of precursors C_m is described by the last term; these neutrons are assumed to be distributed isotropically. Σ_T and Σ_f denote total and fission cross sections, and ν is the number of neutrons produced per fission. χ_P and χ_m are the energy spectra of neutrons produced promptly in fission and by precursor decay, respectively. The probability of a neutron scattering from $dr\, dE'\, d\Omega'$ to $dr\, dE\, d\Omega$ is denoted by Σ_s. The fraction of fission neutrons which are delayed via the formation of a precursor of type m is β_m ($\beta = \sum_{m=1}^{M} \beta_m$), and the rate at which this precursor decays is given by $\lambda_m C_m$. The delayed neutron precursors satisfy the balance equation

$$\frac{\partial C_m}{\partial t}(r, t) = \beta_m \int_0^\infty dE\, \nu\Sigma_f(r, E, t) \int_\Omega d\mathbf{\Omega}\, F(r, E, \mathbf{\Omega}, t) - \lambda_m C_m(r, t), \tag{A.2}$$

$$m = 1, \ldots, M.$$

The scattering nuclei are assumed to be stationary,† and the scattering event is assumed to depend only on the angle between the incident ($\mathbf{\Omega}'$) and emergent ($\mathbf{\Omega}$) neutron directions and not upon these directions. Define

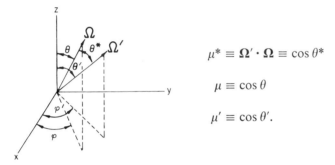

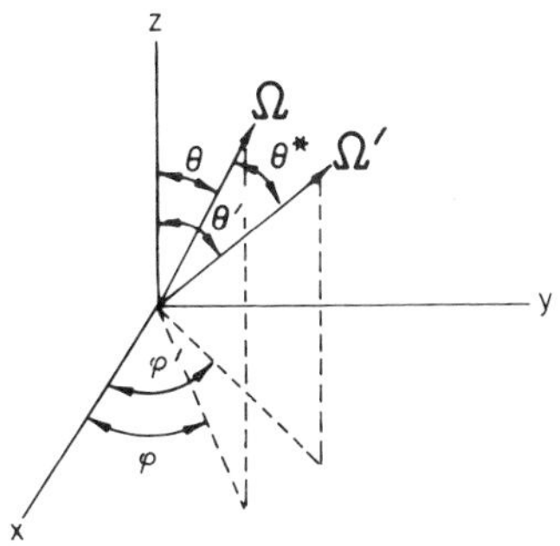

$$\mu^* \equiv \mathbf{\Omega}' \cdot \mathbf{\Omega} \equiv \cos\theta^*$$

$$\mu \equiv \cos\theta$$

$$\mu' \equiv \cos\theta'.$$

Then, the scattering cross section may be expanded,

$$\Sigma_s(r; E', \mathbf{\Omega}' \to E, \mathbf{\Omega}; t) = \sum_{l=0}^{\infty} \frac{2l+1}{4\pi} \Sigma_{sl}(r; E' \to E; t) P_l(\mu^*), \tag{A.3}$$

where P_l is the Legendre polynomial. For elastic scattering, the change in energy is uniquely correlated to the scattering angle by the relation

$$\mu^* = \frac{A+1}{2}\left(\frac{E}{E'}\right)^{1/2} - \frac{A-1}{2}\left(\frac{E'}{E}\right)^{1/2} \equiv \Psi\left(\frac{E'}{E}\right), \tag{A.4}$$

where A is the atomic mass of the scattering nucleus.

Next, the angular dependence of F is expanded,

$$F(r, E, \mathbf{\Omega}, t) = \frac{1}{4\pi}\left[\phi(r, E, t) + 3\mathbf{\Omega} \cdot \mathbf{J}(r, E, t)\right]. \tag{A.5}$$

† Motion of the atomic nuclei is accounted for, without changing the formalism, by defining effective cross sections.

Integrating Equation (A.5) over $\mathbf{\Omega}$ identifies ϕ as the scalar flux,

$$\phi(r, E, t) = \int_{\Omega} d\mathbf{\Omega}\ F(r, E, \mathbf{\Omega}, t),$$

and $\mathbf{J}$ as the current,

$$\mathbf{J}(r, E, t) = \int_{\Omega} d\mathbf{\Omega}\,\mathbf{\Omega}\ F(r, E, \mathbf{\Omega}, t).$$

Equations for ϕ and $\mathbf{J}$ are obtained by substituting Equations (A.3) and (A.5) into Equation (A.1). The resulting equation is then integrated over $\mathbf{\Omega}$, weighted in one case by unity and in the second case by $\mathbf{\Omega}$. The scattering angle is eliminated by using the addition theorem for Legendre polynomials

$$P_l(\mu^*) = P_l(\mu)P_l(\mu') + \sum_{m=1}^{l} \frac{(l-m)!}{(l+m)!} P_l^m(\mu)P_l^m(\mu') \cos m(\varphi - \varphi'), \quad \text{(A.6)}$$

where P_l^m is the associated Legendre function. The resulting equations are

$$\begin{aligned} \frac{1}{v}\frac{\partial\phi}{\partial t}(r, E, t) = &-\nabla\cdot\mathbf{J}(r, E, t) - \Sigma_T(r, E, t)\phi(r, E, t) + Q(r, E, t) \\ &+ (1-\beta)\chi_P(E)\int_0^\infty dE'\ \nu\Sigma_f(r, E', t)\phi(r, E', t) \\ &+ \int_0^\infty dE'\ \Sigma_{s0}(r; E' \to E; t)\phi(r, E', t) \\ &+ \sum_{m=1}^{M} \lambda_m\chi_m(E)C_m(r, t), \end{aligned} \quad \text{(A.7)}$$

$$\begin{aligned} \frac{1}{v}\frac{\partial\mathbf{J}}{\partial t}(r, E, t) = &-\tfrac{1}{3}\nabla\phi(r, E, t) - \Sigma_T(r, E, t)\mathbf{J}(r, E, t) \\ &+ \int_0^\infty dE'\ \Sigma_{s1}(r; E' \to E; t)\mathbf{J}(r, E', t). \end{aligned} \quad \text{(A.8)}$$

A number of further approximations are required to reduce these equations to the familiar diffusion theory. First, the time derivative in Equation (A.8) is eliminated on the basis of the assumption that the rate of change of the current is negligible. Next, inelastic scattering is assumed to be isotropic, which means Σ_{s1} describes only elastic scattering, and elastic anisotropic scattering is assumed to take place without a change in neutron energy, so that

$$\int_0^\infty dE'\,\Sigma_{s1}(r; E' \to E; t)\mathbf{J}(r, E', t) \to \bar{\Sigma}_{s1}(r, E, t)\mathbf{J}(r, E, t).$$

Defining the transport cross section

$$\Sigma_{tr}(r, E, t) = \Sigma_T(r, E, t) - \bar{\Sigma}_{s1}(r, E, t), \tag{A.9}$$

and the diffusion coefficient

$$D(r, E, t) = \frac{1}{3\Sigma_{tr}(r, E, t)}, \tag{A.10}$$

Equations (A.8) and (A.9) may be combined to obtain the energy dependent diffusion equation,

$$\begin{aligned}\frac{1}{v}\frac{\partial \phi}{\partial t}(r, E, t) &= \nabla \cdot D(r, E, t)\,\nabla\phi(r, E, t) - \Sigma_T(r, E, t)\phi(r, E, t)\\ &+ Q(r, E, t) + (1-\beta)\chi_P(E)\int_0^\infty dE'\,\nu\Sigma_f(r, E', t)\phi(r, E', t)\\ &+ \int_0^\infty dE'\,\Sigma_{s0}(r; E' \to E; t)\phi(r, E', t)\\ &+ \sum_{m=1}^{M} \lambda_m \chi_m(E) C_m(r, t).\end{aligned} \tag{A.11}$$

The multigroup approximation to the energy dependence is obtained by integrating Equation (A.11) over the energy interval $E_g \leqq E \leqq E_{g+1}$ and defining

$$\phi^g(r, t) \equiv \int_{E_g}^{E_{g+1}} dE\,\phi(r, E, t),$$

$$1/v^g \equiv \int_{E_g}^{E_{g+1}} dE\,(1/v)\frac{\phi(r, E, t)}{\phi^g(r, t)},$$

$$D^g(r, t) \equiv \int_{E_g}^{E_{g+1}} dE\,D(r, E, t)\frac{\phi(r, E, t)}{\phi^g(r, t)},$$

$$\Sigma_T{}^g(r, t) \equiv \int_{E_g}^{E_{g+1}} dE\,\Sigma_T(r, E, t)\frac{\phi(r, E, t)}{\phi^g(r, t)},$$

$$\nu^g\Sigma_f{}^g(r, t) \equiv \int_{E_g}^{E_{g+1}} dE\,\nu\Sigma_f(r, E, t)\frac{\phi(r, E, t)}{\phi^g(r, t)},$$

$$Q^g(r, t) \equiv \int_{E_g}^{E_{g+1}} dE\,Q(r, E, t),$$

$$\chi_P{}^g \equiv \int_{Eg}^{E_{g+1}} dE\, \chi_P(E),$$

$$\chi_m{}^g \equiv \int_{E_g}^{E_{g+1}} dE\, \chi_m(E),$$

$$\Sigma_s^{g'/g}(r, t) \equiv \int_{E_{g'}}^{E_{g'+1}} dE' \int_{E_g}^{E_{g+1}} dE\, \Sigma_{s0}(r; E' \rightarrow E; t)\, \frac{\phi(r, E', t)}{\phi^{g'}(r, t)},$$

$$g = 1, \ldots, G.$$

With these definitions, Equations (A.11) and (A.2) reduce to Equations (1.1) and (1.2) of Chapter 1 when the total cross section is written as a sum of absorption and scattering terms,

$$\Sigma_T{}^g(r, t) = \Sigma_a{}^g(r, t) + \sum_{g'=1}^{G} \Sigma_s^{g/g'}(r, t),$$

and the scattering removal cross section is defined:

$$\Sigma_s{}^g(r, t) \equiv \sum_{g' \neq g}^{G} \Sigma_s^{g/g'}(r, t).$$

Thus, the quantity obtained by solution of the multigroup diffusion equations is the flux integral over the group, ϕ^g. Information about the energy dependence of the flux within a group, which is required in order to define the group constants, must be obtained elsewhere.

Author Index

Numbers in parentheses are reference numbers and indicate that an author's work is referred to, although his name is not cited in the text. Numbers in italics show the pages on which the complete reference is listed.

Subject Index